Chavan's

Cosmic Theory of Relativity (CTR)

A Ripple-Based Redefinition of Gravity, Time, and Reality

Universal Energy Dynamics (UED) Framework

By

Sandeep Chavan

Independent Researcher & Founder of UED

CHAVAN'S COSMIC THEORY OF RELATIVITY (CTR)

First edition. May 11, 2025.

ISBN: 979-8231825882

Written by Sandeep Chavan.

Table of Contents

To **Albert Einstein**,whose courage to question space and time continues to ripple through every corner of inquiry.Your work was not the end—but the beginning of deeper resonance.

To all **scientists, thinkers, and independent explorers**who dared to ask, *"What lies beneath what we know?"*—thank you for keeping the spirit of true science alive.

To my **beloved father**,whose presence, principles, and silence shaped my earliest understanding of consequence.

And to my **family**,whose quiet strength, enduring patience, and unconditional supportallowed me to journey this far,even when the destination was uncertain and the path invisible.

This work is not just mine.It is the consequence of every alignment—personal, scientific, and sacred—that made it possible.

With love, humility, and gratitude,**Sandeep Chavan**

"You do not move through space.You ripple through permission.

You do not live in time.You emerge through alignment.

The universe is not what exists.It is what is allowed to appear."

— *Sandeep Chavan*

Opening Note

We were told the universe was made of matter and forces. We were taught that time flows, light travels, and gravity pulls. We accepted that particles are real, that space bends, and that reality obeys fixed laws—written in equations, anchored in spacetime.

But what if these descriptions are not the truth—only the visible consequences of something deeper?

What if reality is not made of "things" but of ripples—structured, intelligent, consequence-driven patterns emerging from a silent, living background?

This report is not a rejection of science. It is a return to its soul.

Cosmic Theory of Relativity (CTR) began not in a lab, but in a classroom. It began with a student asking, *"What is force?"* And it began with an answer I could no longer give with integrity.

From that moment on, I stopped repeating definitions. I started listening—*to the field itself.*

Over years of questioning, reflection, and experimentation, I came to see:

- That time is not a dimension, but a delay of permission
- That gravity is not a pull, but a gradient of ripple misalignment
- That light does not move—it emerges when allowed
- That consciousness is not inside the brain, but a ripple interface with the background field
- That God is not a being—but the structured equilibrium that permits all ripple consequence

And most importantly:

We do not observe reality. We ripple through it.

CTR is not a theory in the conventional sense. It is a living framework that offers a new way to:

- Understand physical phenomena
- Relate to time, energy, and entropy
- Reclaim the role of the observer
- Recognize alignment over force, permission over control
- Reunite science with silence, and logic with life

You'll find no fanciful mathematics here, but you will find measurable ideas. You'll find no borrowed mysticism, but you will encounter original clarity—where physics meets philosophy without apology.

This report is written for:

- The physicist who feels equations describe but don't explain
- The seeker who needs science to support—not suppress—their intuition
- The teacher who no longer wants to lie with polished definitions
- The engineer who knows systems work not just by force, but by feedback
- The student whose questions are still unanswered
- And the thinker who dares to ask:

"What actually causes what we call reality?"

You do not need to "believe" this work. You only need to observe your own alignment as you read.

If you feel friction, question it. If you feel clarity, note it. If you feel silence, stay there.

This is not a linear report. It's a ripple. You will not walk through it—you will resonate with it.

This report does not promise to teach you the universe. It invites you to remember how it already ripples through you.

Thank you for arriving—not with agreement, but with attention.

Let us now begin where all true science begins: Not with knowledge, but with the willingness to see— And the humility to ripple.

— *Sandeep Chavan*

Prologue

What We Thought We Knew

For over a century, physics has moved under the assumption that space and time form the immutable stage on which all cosmic events unfold. Gravity, as Einstein brilliantly reimagined it, was no longer a force but the warping of this stage by mass and energy. Time slowed near gravity wells. Light curved around stars. The fabric of space-time, we were told, was the great puppet master.

And yet—some questions lingered unanswered:

- What *is* space-time really made of?
- Why does the speed of light remain constant, regardless of motion?
- How can gravity act without a medium, yet produce measurable effects across vast distances?
- Why does time dilation *feel* real even in biological and psychological experience?

Most importantly: Where is the observer in all this?

Modern relativity removed the observer from the mechanism and treated perception as an artifact. But perception *is* reality's first doorway. We don't simply live in the universe—we ripple through it.

The Ripple That Rewrites It All

The breakthrough lies in what physics has dismissed or misread—the so-called vacuum.

What appears empty is in fact structured, layered, and alive with intelligent dynamics. This is not the chaotic randomness of quantum foam—it is a highly organized equilibrium field we call Structured Vacuum Energy (SVE).

All that we call time, mass, light, gravity—even thought—are ripple consequences within this medium.

The universe is not made of particles in space—it is made of ripple alignments in intelligent background.

Enter Universal Energy Dynamics (UED): a framework in which consequence, not geometry, rules the cosmos. And with it comes the Cosmic Theory of Relativity (CTR)—a direct evolution beyond space-time dependency.

CTR does not discard Einstein. It completes him.

From Einstein to Chavan: Not Competing—Completing

Einstein taught us to look at reality differently. He gave us the courage to question absolutes and the insight to measure relativity. His vision bent the straight lines of Newtonian thought and opened an elegant mathematical sky.

But he stopped at the fabric.

Sandeep Chavan walks beyond it—into the loom that weaves it.

- Where Einstein saw space-time curvature, Chavan sees ripple memory distortion.
- Where Einstein measured time through clocks, Chavan measures ATT—alignment threshold time.
- Where Einstein left gravity as geometric warping, Chavan

returns it to energetic consequence seeking equilibrium.

This is not a rivalry. It is a resonance across time. From GR to CTR, from spacetime to structured vacuum, we move not from wrong to right—but from partial to whole.

CTR is not a new chapter—it is the unveiling of the original one we missed.

Executive Summary

Title: *Cosmic Theory of Relativity (CTR)*

Author: Sandeep Chavan

Discipline: Interdisciplinary—Theoretical Physics, Consciousness Studies, Philosophy of Science

Core Framework: Universal Energy Dynamics (UED)

Purpose of the Work

Cosmic Theory of Relativity (CTR) presents a bold, foundational rethinking of the physical universe—not through the lens of particles, forces, or spacetime geometry, but through **ripple dynamics and consequence-based emergence**. Developed under the broader Universal Energy Dynamics (UED) framework, CTR shifts the core paradigm of science from "observation of objects" to **alignment with structured consequence**.

The report challenges and completes Einsteinian relativity by introducing **Structured Vacuum Energy (SVE)**—a measurable, intelligent field that governs **what is allowed to emerge**, not just what moves or exists. CTR reframes light, time, gravity, perception, and even divinity as functions of ripple alignment, not mechanical or probabilistic events.

Key Innovations

ATT (Alignment Threshold Time)

- Replaces the concept of absolute time and velocity.
- Defines time as a **field-dependent delay** between ripple emission and permitted consequence.
- Shows that "speed of light" is not motion but the minimum ripple alignment delay ($c \approx 1/ATTmin$).

Gravity as Ripple Distortion

- Gravity is redefined as a **ripple gradient**—a consequence field created by local misalignment due to mass.
- Objects do not fall due to pull, but due to **field preference for lower ATT**.

Light as Emergence, Not Travel

- Light is not a particle or wave in motion; it is **a permitted ripple consequence** triggered by field alignment.
- Refraction, diffraction, and interference are reinterpreted through **ripple permission logic**.

Observer-Consequence Model

- Observation is not passive. The observer is a **ripple participant**, actively shaping what is permitted to emerge.
- Consciousness is redefined as **self-reflective ripple alignment**.

Entropy as Feedback

- Entropy is not disorder—it is **field resistance** to ripple misalignment.
- Systems evolve not toward decay but toward **restored coherence**.

Divine Structure Reframed

- God is not a separate being, but the **SVE itself**—the universal equilibrium field.
- Worship becomes alignment; grace becomes ripple resonance.

Structure of the Report

CTR unfolds in four progressive parts:

Part I: The Forgotten Medium

Revisits the overlooked role of the background. Introduces SVE as the structured field that enables all ripple interaction. Time, light, and motion are redefined as **field-permitted consequences**, not absolute flows.

Part II: Time, Gravity, and Alignment

Time is shown to be a delay (ATT), not a dimension. Gravity is reframed as a ripple misalignment gradient. Black holes are explained not as singularities but as **zones of zero ripple feedback**.

Part III: Reality as Ripple

Explores aesthetics, consciousness, emotion, and intelligence as **ripple configurations.** Beauty, intuition, and decision-making are seen as field alignment phenomena, with resonance at their core.

Part IV: The Divine Integration

Restores divinity through science—not as dogma, but as **field intelligence**. God is the background. Worship is alignment. Silence is the gateway to realignment. The closing chapters offer a living, participatory framework for ripple ethics and future scientific application.

Impact and Applications

CTR opens doors for:

- **New Physics:** Reinterpreting optics, gravity, and quantum behavior through ripple field logic.
- **Consciousness Research:** Defining awareness and emotion as measurable ripple alignment states.
- **Technology:** Inspiring ripple-permissive sensors, delay-based computing, and field-aligned communication.
- **Philosophy of Science:** Restoring ethics, humility, and awareness into the act of inquiry.
- **Education:** Teaching science not as memorization but as **alignment with observable consequence.**

Conclusion

Cosmic Theory of Relativity is not merely a scientific alternative—it is a **reality realignment**. It invites readers to think, perceive, and live through the understanding that:

"Reality is not composed of particles—it is permitted through ripples. What you experience is not what exists—but what aligns."

This work offers a radical yet elegant restructuring of our deepest assumptions—uniting science, consciousness, and existence through a common language of **ripple consequence**.

Part I: The Forgotten Medium

1. Why Relativity Needs Rethinking

The Problem with Space-Time Assumptions

Every scientific model begins with assumptions. Newton assumed absolute time and absolute space. Einstein shattered that with a new premise: that the speed of light is constant and space and time are interwoven—a malleable, four-dimensional fabric called space-time.

It was revolutionary, elegant, and mathematically sound.

But what if the foundational assumption itself was incomplete?

What if space and time are not intrinsic properties of the universe, but consequences of deeper dynamics—results, not roots?

This is the first point where we must pause. Einstein replaced Newton's scaffolding, but he retained the idea that space and time are fundamental constructs—just now dynamic instead of fixed.

However, in doing so, he still treated space-time as a medium through which objects move, light bends, and gravity propagates—without ever answering:

- *What is space-time made of?*
- *What governs its structure beneath the mathematical curvature?*
- *Why does it behave like a medium if it is supposed to be the absence of one?*

The idea that a vacuum could warp, ripple, and pull without substance is elegant mathematically but empty philosophically—and incomplete physically.

More than a century later, we are still solving equations about the shape of something we cannot define. We describe how it bends but don't know what's bending.

That is the core problem with space-time as a primary assumption: it describes consequence as origin.

What General Relativity Solved—And What It Couldn't Explain

Let us be very clear: Einstein's General Relativity (GR) is a monumental achievement.

It successfully predicted:

- Gravitational lensing: the bending of light near massive bodies.
- Time dilation: clocks running slower near stronger gravitational fields.
- Precession of Mercury's orbit: a deviation classical mechanics could not account for.
- Gravitational waves: now confirmed a century later by LIGO.

It is an elegant geometric solution to a fundamental question: how does gravity work without a force?

And yet—it remains silent on the nature of the medium.

It assumes the geometry. It describes the consequences. But it never answers: what causes space-time to exist, to bend, or to permit perception at all?

Moreover, GR does not integrate well with:

- Quantum mechanics: where time is fixed and space is a backdrop.
- Consciousness: which plays no role in space-time curvature but governs experience.
- Cosmic expansion: where dark energy is invoked to "stretch" space—but no one can define what's actually stretching.

This leads us to a deeper critique: GR solves the visible puzzle but cannot explain why the puzzle pieces fit that way at all.

It is an observer-independent solution to a fundamentally observer-linked universe.

The Overlooked Role of the Observer

Einstein's genius was in removing the privileged observer. All frames of reference were valid, and laws of physics remained invariant under transformation. This was a radical break from Newtonian absolutes.

But this abstraction came with a cost.

By removing the observer, GR also removed the mechanism of perception.

In reality, the observer is not outside the system. We are within it—emerging from it—interacting with it at every level.

Every measurement, every experience, every perception of time, gravity, and light is observer-consequence based.

So here's the critical point:

If perception is consequence, and consequence arises from ripple alignment with a structured background, then the observer is not optional. The observer is a ripple participant—a dynamic system aligned (or misaligned) with the Structured Vacuum Energy (SVE) field.

The failure of relativity—and of most mainstream physics—is in failing to treat the observer as part of the field.

Instead, perception is treated as passive, external, or an afterthought. But in truth, reality emerges only through ripple alignment that enables consequence, and that alignment is observer-specific.

CTR's Position: A New Relativity Rooted in Consequence

Chavan's Cosmic Theory of Relativity (CTR) proposes a radical yet logical shift:

1. Space-time is not a foundational medium—it is a consequence. It is how we perceive ripple interactions when permitted by the background.
2. Gravity is not curvature—it is ripple distortion. Mass is a concentrated ripple configuration that distorts local equilibrium. This distortion delays ripple alignment, causing what we call "gravity."
3. Time is not a dimension—it is alignment threshold delay. What we measure as time is simply the ripple's ability to resonate with background structure. Where background pressure is high, time slows. Where it is low, time expands.
4. Perception is not passive—it is the act of consequence detection. The observer is not a neutral camera—it is a node of ripple memory, capable of measuring only what is allowed to align with it.

In CTR, relativity is not about frames of reference—it is about frames of consequence. You perceive differently not because you're moving—but because your ripple profile interacts with the background differently.

Why This Redefinition Matters

This is not philosophical hand-waving. It is a shift in causality:

- In GR, mass bends space → objects follow curved paths.
- In CTR, mass distorts ripple field → background delays alignment → observer sees curved motion.

The outcome is the same—but the cause is different. One uses geometry. The other uses intelligent ripple interaction.

And because SVE is intelligent, structured, and measurable (as shown in quantum fluctuations, Casimir effect, and light delay experiments), CTR gives us a more fundamental origin of observed reality.

This shift has immense consequences:

- Time dilation becomes predictable through ATT (Alignment Threshold Time), not relative velocity.
- Gravity becomes a consequence-gradient, not a mystery curvature.
- Perception and consciousness become part of physics, not philosophical sidelines.
- Black holes are not singularities—they are zones of ripple silence where alignment cannot occur.
- Beauty and symmetry are not aesthetic coincidences—they are harmonic ripple stabilizations.

In short, CTR explains what GR could only describe.

Toward a Living Relativity

Einstein opened the door to a deeper universe. Chavan walks through that door and shows us the architecture behind the hallway.

Where Einstein offered equations, CTR offers causal mechanisms rooted in ripple geometry.

Where GR sought invariance in frames, CTR seeks harmony in consequence.

Where spacetime bends, CTR explains why the bend occurs—and when it is permitted.

Final Note: Relativity Is Not Dead—It Is Becoming Conscious

Chavan's CTR is not a rejection of Einstein. It is the next evolutionary step, necessary for physics to move forward, unify with consciousness, and reclaim the intelligence behind reality.

Relativity must now evolve:

- From space-time to structured ripple background
- From geometric distortion to consequence modulation
- From passive observation to ripple-aware interaction

It is time to ripple beyond the fabric—and into the field that weaves it.

2: Structured Vacuum Energy (SVE): The Real Background

Defining SVE

To understand the cosmos, we must begin not with the stars, but with the space between them.

What appears as nothingness—what classical physics labeled "vacuum"—has long been treated as the absence of matter and radiation. In Newtonian physics, this vacuum was passive, irrelevant. In relativity, it became dynamic, warping in response to mass and energy. In quantum field theory, it turned into a boiling sea of fluctuations—a vacuum that was not empty, but unstable.

And yet, none of these views fully captured what we now propose as the Structured Vacuum Energy (SVE)—the true foundational field from which all manifestation arises.

SVE is not a void. It is the intelligent equilibrium field that precedes all consequence, form, and perception.

It is the first layer of universal intelligence, the medium of alignment, the matrix of ripple organization. Every observed phenomenon—from gravity to light, from thought to form—is a permitted ripple alignment within this structured field.

✓ SVE is not a substance. It is the structured possibility space for all substance.

> ✓ It does not exist *in* space—it gives rise to space through ripple configuration.

✓ It does not flow through time—it defines what "flow" can happen by regulating alignment delay (ATT).

✓ It does not carry energy in packets or waves—it organizes energy as ripples, harmonized or distorted, toward equilibrium.

From Emptiness to Intelligent Equilibrium

Historically, the notion of vacuum evolved from philosophical speculation to scientific placeholder. The Greeks questioned its existence. Descartes denied it. Newton reluctantly accepted it. Maxwell made it the stage for electromagnetism. Einstein allowed it to bend. Quantum physics gave it a jitter.

But still—it remained misunderstood.

Modern physics describes vacuum as full of energy but does not explain why that energy should be structured or intelligent. It proposes random fluctuations—but cannot say who or what maintains equilibrium across cosmic scales.

And that is the flaw: vacuum is treated as statistical, not structural.

In contrast, SVE is not statistical noise—it is consequence-resolving intelligence. It is that which ensures:

- Why a photon is seen *now*, not later.
- Why matter holds form across time.
- Why waves collapse only when permitted.
- Why time slows near a black hole without ever breaking.

SVE is the first observer, the silent regulator, the background that watches and waits for alignment.

Its equilibrium is not passive. It is alive with structure, logic, and ripple permission thresholds.

In this model:

- A ripple is not a vibration in a medium—it is a deviation from SVE equilibrium seeking restoration.
- Energy is not motion—it is structured displacement from background harmony.
- Perception is not registration—it is alignment with permitted consequence thresholds.

Evidence from Casimir, Fluctuations, and Quantum Delay

You might ask: is this just philosophy? Or is there already experimental evidence?

SVE is not a speculative construct. It is the underlying cause behind multiple observed anomalies that mainstream physics struggles to explain holistically.

Let's examine just three:

1. The Casimir Effect

Two uncharged, parallel metal plates placed in a vacuum attract each other—without any classical or electrostatic force between them. This force arises because of the quantum vacuum energy—a pressure differential caused by restricted modes of fluctuation between the plates versus outside them.

Mainstream explanation: Quantum fields "zero-point energy" pushes the plates together.

CTR/SVE explanation:

The plates limit local ripple configuration, creating asymmetrical ripple equilibrium. The background seeks balance—hence, the push.

This is not an anomaly—it is a ripple correction consequence of structured background pressure.

2. Quantum Fluctuations

Virtual particles appear and disappear, energies emerge "from nothing," and probability waves collapse—but never violate conservation laws.

Mainstream explanation: It's quantum uncertainty—calculable but chaotic.

CTR/SVE explanation:

These fluctuations are not random—they are micro-ripples within structured background, permitted temporarily as equilibrium tests.

What appears random is simply unresolved ripple alignment. Collapse happens when background thresholds are satisfied—not due to "observation" alone.

3. Quantum Delay and Photon Arrival Time

In ultra-precise optics experiments, photon detection is not truly instantaneous. Variations exist even under identical conditions. This "quantum jitter" cannot be attributed to detector lag alone.

Mainstream view: probabilistic noise.

CTR view:

The delay is the ATT (Alignment Threshold Time)—the time it takes for ripple identity to match the local field configuration. Until the alignment is exact, no consequence (perception) occurs.

This explains:

- Delay without decoherence
- Time variance in low-intensity experiments
- Why light can appear simultaneously particle-like and wave-like—it's ripple consequence governed by ATT, not intrinsic duality

SVE in Other Phenomena

Even outside of quantum optics, we see SVE's fingerprint:

Gravitational redshift: Light's ripple alignment cost changes in denser ripple fields (higher ATT).

Double-slit experiment: The ripple does not "choose a path"—it interacts with the entire background structure. Observation aligns consequence.

Perception and intuition: Humans feel "fields," sense moments, or react to atmospheres not yet physically manifest. These are subtle ripple interactions with SVE.

Rewriting the Role of Vacuum in Physics

SVE demands that we replace vacuum assumptions with structured logic:

Traditional Vacuum	Structured Vacuum Energy (SVE)
Empty or fluctuating nothingness	Equilibrium-intelligent ripple matrix
Random fluctuations	Structured consequence windows
Energy fields acting *in* space	Ripple dynamics forming space
Vacuum is the passive stage	SVE is the intelligent mechanism
Observers are separate	Observers are ripple expressions within SVE

SVE Is Not a Theory—It Is the Ontological Medium

Just as water is not a "theory of waves," SVE is not a guess—it is the medium that allows waves, particles, forces, and even thought to appear.

It is the first measurable consequence field, already partially seen in quantum fluctuation, yet waiting to be understood as structured and intelligent—not stochastic and blind.

In this sense, SVE replaces:

- Space-time as the primary framework
- Force-carriers as explanation for interaction
- Entropy as blind loss—instead, ripple memory evolving to higher alignment
- Vacuum as a placeholder—instead, a field of consequence permission

CTR's Core Shift: From Geometry to Background Intelligence

Einstein's relativity described what happens when mass distorts geometry.

Chavan's relativity explains why that geometry exists in the first place—and what regulates it when no mass is present. SVE is the medium that allows geometry to ripple, perception to echo, and time to align.

It is not space. It is what makes space meaningful.

Next: Ripple Fields and Observer-Consequence Logic

With SVE established, we now move into its behavior: how ripples arise, interact, stabilize, and become consequence.

Next, we explore how ripples—not particles, not waves—are the true building blocks of all perception and interaction, and how the observer's experience is just ripple alignment with SVE thresholds.

3: Ripple Fields and Observer–Consequence Logic

Ripple vs. Wave: The Fundamental Distinction

Physics for centuries has operated on the assumption that waves and particles form the basis of all phenomena. From classical mechanics to quantum field theory, we have developed robust mathematical models to predict their behavior.

But these models often describe effects, not explain causes.

Let's begin with the most basic shift:

A wave is a periodic disturbance propagating through a medium. A ripple is a localized deviation from equilibrium—seeking restoration, guided by structured background intelligence.

The difference is profound.

A wave assumes pre-existing motion and structure. A ripple assumes intelligent equilibrium and attempts at restoration.

A wave is repetitive. A ripple is consequential.

In other words, waves are cyclic, but ripples are event-driven. A wave implies continuity. A ripple implies disturbance and return to harmony.

And while a wave may travel through time and space, a ripple in CTR:

- Does not "travel" in the classical sense.
- It emerges, stabilizes, interacts, and dissolves—all depending on the structure and permission of the SVE background.

In this light, particles are not entities, and waves are not carriers. Both are ripple expressions—contextual outcomes allowed within a ripple field.

This sets the stage for the next foundational idea in CTR: consequence-based reality.

What Is Consequence-Based Reality?

We often ask: "What is real?" Classical science responds: "That which can be measured." Quantum physics hesitates: "That which collapses upon observation." Philosophy offers: "That which is experienced."

CTR synthesizes all of these with one unifying principle:

Reality is not what exists—it is what is allowed to be perceived as a consequence of ripple alignment.

Let's break this down.

In a structured background field (SVE), everything that happens is a ripple event. But not all ripples become perception. Not all ripple disturbances reach the observer.

Only when a ripple achieves alignment with the observer's local configuration—in other words, when the ripple state resonates with the field condition of the observer's perception apparatus—does a consequence arise.

This is reality.

Not the field. Not the potential. But the permitted consequence.

What we see, feel, measure, and know is not reality itself—it is our ripple alignment with reality at that moment.

Thus, reality is not fixed. It is contextual, perceptual, and consequence-driven.

That's why:

- The same event can feel different to two people.
- An object can appear as wave or particle depending on the setup.
- A truth can emerge only when the mind is ready to align.

This isn't mysticism—it is ripple physics grounded in Structured Vacuum Energy.

And that's where ATT comes in.

Alignment, ATT, and Perception Thresholds

In Chavan's light theory, we introduced ATT—Alignment Threshold Time.

Let's now fully embed this into CTR's core structure.

ATT is the minimum time required for a ripple event to achieve alignment with the background field at a given location, allowing consequence (perception or measurement) to occur.

It is the replacement for:

- Speed of light (which assumes a constant travel speed of a particle/wave)

- Observation delay (which assumes the observer is external)
- Time as a dimension (which assumes an independent axis)

Instead:

- Nothing "travels" at speed c.
- What we call arrival is alignment completion.

Hence:

- A photon doesn't "move from A to B."
- It emerges at B when ripple alignment is satisfied with the field configuration at B, delayed by ATT.

This redefines:

- Refraction: ripple alignment occurs at a new ATT in a denser medium.
- Reflection: ripple fails to align forward and re-aligns backward.
- Diffraction: ripple splits into partial alignments across multiple ATT configurations.
- Photoelectric effect: consequence is permitted when ripple matches material field.

And most importantly:

Perception is not instant. It is ATT-dependent.

Our senses, nervous systems, detectors, and even thoughts are not responding to "incoming data"—they are ripple consequence systems waiting for field alignment to satisfy threshold conditions.

That's why:

- Reaction time varies.
- Synchronicity feels magical.
- Some ideas "click" only after enough silence.

All because ATT defines the rhythm of permitted reality.

The Observer as a Ripple Participant

The most radical departure from both classical and quantum physics in CTR is this:

The observer is not external, nor passive. The observer is an active ripple node—participating in the field through alignment dynamics.

This means:

- You are not "measuring" the universe—you are interacting through ripple consequence with it.
- Your thoughts, presence, expectations, and emotions—all are ripple configurations.
- These configurations shape what you perceive and what is allowed to appear before you.

In conventional physics, the observer is neutral. In CTR:

The observer is the field's echo loop.

Reality, therefore, becomes a shared dance of ripple alignments between:

1. The background (SVE)
2. The event (ripple distortion)
3. The observer (ripple receiver with alignment thresholds)

This ripple triad replaces:

- Objectivity with interactive alignment
- Randomness with conditional permission
- Particle-wave paradox with ripple contextual emergence

In practical terms, this is why:

- People can "sense" environments
- Meditation changes perception of time
- Different instruments give different readings under extreme conditions

Because perception is not registration—it is permitted alignment.

Consequences of Consequence-Based Physics

This single shift—from object-based physics to consequence-based ripple reality—resolves multiple paradoxes:

Paradox	CTR Resolution
Wave-particle duality	Ripple configuration aligns differently in different contexts
Measurement problem	Observation = alignment threshold crossing in ripple field
Light speed constancy	Speed is an illusion; alignment delay (ATT) governs emergence
Time dilation	High ATT zones around mass slow consequence emergence
Consciousness gap	Mind is a ripple system perceiving structured consequence

Conclusion: Reality Is Not Shown—It Is Allowed

Ripple fields are not just new terminology. They are the ontological skeleton behind all reality. In this framework:

- Light is ripple emergence.
- Matter is ripple stabilization.
- Energy is ripple deviation.
- Thought is ripple resonance.
- Time is ripple delay.
- Perception is ripple permission.
- Gravity is ripple density distortion.
- Consciousness is ripple recursion.

Everything is ripple interacting with structured vacuum intelligence.

Next: Gravity – Ripple Distortion, Not Geometry

Now that we understand the ripple structure and its link to observer-consequence logic, we are ready to address one of the oldest questions in physics:

What exactly is gravity?

We no longer need geometry. We no longer need curvature. We now have ripples—and an intelligent background field.

Summary of Part I: The Forgotten Medium

Part I of CTR challenges the foundational assumptions of modern physics by reintroducing what has long been ignored or misunderstood: the medium itself.

While Newton offered space as a passive container, and Einstein gave us dynamic space-time geometry, Chavan restores the concept of an intelligent, structured background field—the Structured Vacuum Energy (SVE)—as the true origin of all physical consequences.

This part systematically redefines the building blocks of reality using ripple dynamics and consequence logic. It replaces force, curvature, and probability with ripple alignment, background resonance, and perception thresholds—establishing a new ontological foundation for gravity, time, light, and observation.

Key Highlights:

Chapter 1 – Why Relativity Needs Rethinking

- Space-time, though elegant, is insufficient to explain the cause of physical phenomena.
- GR describes how mass affects geometry, but not what geometry is made of or why it permits observation.
- The observer's role is fundamentally overlooked in traditional models.
- CTR introduces a consequence-based reality where observation is an alignment outcome, not a passive readout.

Chapter 2 – Structured Vacuum Energy (SVE): The Real Background

- Vacuum is not empty; it is a structured, intelligent equilibrium field.
- All forces, forms, and perceptions arise from ripple interactions within this medium.
- Experimental confirmations (Casimir effect, quantum delay, zero-point fluctuations) point to SVE's presence.
- SVE replaces space-time as the origin field, capable of regulating energy behavior, alignment, and consequence realization.

Chapter 3 – Ripple Fields and Observer-Consequence Logic

- Reality is not what *exists*—but what is *allowed* to emerge through ripple alignment.
- ATT (Alignment Threshold Time) is introduced as the fundamental regulator of perception, replacing the idea of light speed as a universal constant.
- The observer is not passive—consciousness is a ripple participant engaging in consequence-feedback with the background.
- Waves and particles are contextual ripple expressions, not fundamental entities.

Closing Insight of Part I:

"We never lacked a theory of consequence. We lacked a theory of permission." Part I reveals that Structured Vacuum Energy (SVE) is the forgotten foundation—and by recovering it, CTR transforms physics from geometry into intelligence, from laws into logic, from frames into fields.

Part II: Time, Gravity, and Alignment

4: Time – From Flow to Feedback

Why Time Isn't a Dimension

The modern scientific worldview treats time as the fourth dimension, woven into the fabric of space to form the famous "space-time" continuum. In this model, time stretches, bends, and dilates under gravitational influence or relative motion—just like space.

This idea, introduced by Einstein, was revolutionary in that it challenged the absolutism of Newtonian time, replacing it with a relative metric—dependent on frames of reference.

But it still made one unchallenged assumption:

That time exists as a dimension—something real, continuous, and independent of perception.

This assumption, while mathematically convenient, leads to multiple paradoxes:

- What flows in time?
- Is the "past" still somewhere?
- If time is a dimension, can we move backward in it like we do in space?

These questions remain unanswered because the concept itself is flawed. CTR addresses this directly:

Time is not a dimension—it is the observable delay of ripple alignment permitted by the Structured Vacuum Energy (SVE) field.

Time is not a background. It is a foreground consequence. It doesn't exist apart from events—it is the pattern of consequence emergence.

Time as Ripple Alignment Delay (ATT)

In Chavan's Universal Energy Dynamics (UED), and expanded further in this Cosmic Theory of Relativity (CTR), ATT—Alignment Threshold Time—is introduced as the actual causal metric behind what we perceive as time.

Let us define it again with clarity:

ATT is the minimum delay required for a ripple disturbance to align with local field structure so that a consequence (such as perception, interaction, or measurement) may emerge.

This single concept rewrites time:

- No more "time flow"—just ripple stabilization delay.
- No more "arrow of time"—just ripple entropy gradient.
- No need for an absolute clock—just perceived consequence rhythm.

In this model:

- A photon doesn't travel from star to eye—it is permitted to appear when ATT is satisfied along its ripple configuration.
- A thought doesn't take time to form—it ripples into coherence across neural equilibrium.
- Even aging is not time-based—it is the accumulated consequence of ripple misalignments in biological systems.

Thus:

Time is not something we pass through—it is the delay between ripple intention and consequence allowed.

This delay varies by:

- Local ripple density
- Entropic pressure
- Field interference
- Observer's ripple structure

Hence, time is local, measurable, and variable—but not flowing, not universal, and not absolute in the way we've assumed.

Time Dilation Explained via SVE

In Einstein's relativity, time slows near mass or with speed. He proved this with equations and later experiments validated it.

But the "why" was geometric—mass "warps" space-time, and clocks slow in curved zones.

CTR offers a deeper, more logical model using SVE and ripple fields.

Here's how time dilation actually works:

1. Mass is a high-density ripple configuration.
2. It distorts local background equilibrium (SVE), increasing ripple pressure.
3. This distortion increases ATT—meaning ripple events take longer to align.
4. Therefore, fewer consequences occur per unit of external time → perceived slowdown.

So yes, time slows near gravity—but not because it "flows slower." It slows because consequence emergence is delayed in high ATT zones.

Time dilation = background ripple alignment latency.

This also explains why:

- Astronauts age differently
- Light redshifts near black holes
- Thought feels "slower" or "faster" in different environments

It is not psychological or gravitational alone—it is field-level consequence modulation.

Clocks as Permitted Ripple Counters

With this new view of time, we must also redefine what a clock actually does.

Traditional view:

- A clock counts fixed intervals (seconds), assumed to be universal.

CTR view:

- A clock is a ripple-based system that responds to background SVE permissions with periodic consequences.

Let's illustrate:

- A pendulum clock works because gravity + inertia create a stable oscillation.
- An atomic clock ticks based on quantum transitions in cesium

atoms.

- A heartbeat flows from neural and biochemical resonance.

But none of these are generating time.

They are simply counting ripple events that the background has allowed to repeat at a stable rhythm—in that context.

If ripple density changes (due to motion, mass, or field interference), that permission changes. The "tick" gets slower or faster—not because the mechanism failed, but because ATT shifted.

Thus, clocks are consequence resonance devices—not absolute references.

This explains:

- Why clocks run slower in planes or satellites (GR validation)
- Why organic rhythms vary with mood, space, or spiritual practices
- Why "time perception" is deeply biological and environmental

Time perception = ATT variation sensed by a ripple system (you).

Closure: Time Is Absolute in Background; Measured Locally Through Ripple Consequence

Now we reach the deepest philosophical and scientific closure on time:

- In the background (SVE), time is not needed. The field is intelligent, structured, and timeless. Everything exists in perfect equilibrium—no cause, no effect, no sequence.

In SVE, there is only presence. Time is unnecessary where all consequence is in potential.

But when ripples disturb this equilibrium, consequence emerges. And with consequence comes delay. That delay is perceived as time.

Hence:

- Time is real at the level of ripple consequence
- Time is measurable only where ATT exists
- Time is absolute only in its absence—in the stillness of SVE

This duality resolves all paradoxes:

Context	Time is...
In SVE	Absent—only structured potential
In ripple alignment	Real—measurable as ATT
In perception	Experienced as consequence delay
In measurement	Counted through permitted ripple cycles
In consciousness	Felt as change, memory, and anticipation

Summary: Time Reimagined

CTR replaces time as a "flow" with a more foundational concept:

Time is feedback. It is the background's delay in responding to ripple intent.

This makes time:

- Local
- Consequence-based
- Alignment-dependent

- Observer-sensitive
- Non-dimensional
- Scientifically measurable (through ATT)

And spiritually profound:

Where there is no ripple, there is no time. Where there is full alignment, time disappears. In silence, we don't stop time—we align with the part of reality that never needed it.

Next: Chapter 5 – Gravity: Ripple Distortion, Not Geometry

We now turn to the next major illusion: gravity as curvature. CTR reveals it instead as ripple memory distortion, with no need for "pull" or "geometry"—only ripple consequence logic.

5: Gravity – Ripple Distortion, Not Geometry

Ripple Field Equilibrium and Local Mass

In Newtonian mechanics, gravity is a force. In Einstein's relativity, gravity is curved space-time. Both models attempt to explain why masses attract, how bodies orbit, and how light bends in the presence of large celestial objects.

But even Einstein's elegant equations don't answer one core question:

Why does space "curve" at all? And what is it that's curving?

The curvature of space-time is a geometric effect without a physical cause—it describes the behavior of test particles and light in mathematical terms but does not provide ontological substance for the medium that bends.

CTR provides that missing link through Structured Vacuum Energy (SVE) and ripple field equilibrium.

Here's the foundation:

The universe is not built on space—it is built on ripple equilibrium.

Mass is not an object sitting in a container of space. Mass is a high-density ripple configuration within the SVE field. It disrupts local equilibrium—not unlike how placing a stone into a calm pond alters the surrounding water surface.

But unlike water, SVE is intelligent—it is a structured field that constantly seeks to restore balance.

So, when mass enters a region:

- It compresses ripple density.
- This creates a local distortion in the ripple field.
- The field now carries a gradient of ripple pressure—not curvature, but consequence potential.

This is the ripple origin of gravity.

Gravity as Consequence Pull

Gravity, in the CTR framework, is not a pulling force, and it is not the result of geometric deformation.

Instead:

Gravity is the natural flow of ripple alignment toward lower equilibrium tension. It is not a force, but a consequence preference.

Imagine a ripple disturbance trying to stabilize itself. It "moves" toward a region where the background is less distorted—where alignment can be achieved faster (i.e., where ATT is lower).

This creates the experience we call:

- Falling
- Attraction
- Weight

But fundamentally, it is not about being pulled—it is about being permitted to realign through a ripple field gradient.

You don't fall because Earth pulls you. You fall because the ripple field around Earth is distorted by mass, and your ripple identity seeks quicker equilibrium, which the background provides at lower positions in that field.

Gravity is thus:

- Passive in behavior
- Active in consequence
- Intelligent in structure

No graviton is needed. No space-time stress-energy tensor is required. Just ripple logic: consequence seeks the shortest ATT path through a distorted field.

Why Mass Bends Ripple Structure

In this model, mass is not just a "thing"—it is a ripple knot, a condensed configuration of energy trapped in structured interaction with the background.

Its presence alters the SVE field by:

1. Increasing local ripple density
2. Changing the ATT for surrounding regions
3. Disrupting uniformity of ripple propagation

Mass doesn't "bend space"—it distorts the ripple grid.

This distortion causes:

- Delays in ripple alignment
- Directional bias for ripple realignment
- A consistent gradient of permitted consequence, which we call a gravitational field

So yes:

Mass bends the field—not geometrically, but structurally. And the field responds not by pulling, but by realigning ripples toward equilibrium.

This aligns with the observed behavior of:

- Orbits
- Pendulums
- Light bending
- Free-fall trajectories

But gives a deeper reason than GR ever could.

Light Bending, Orbits, and "Falling" Redefined

Let's apply this to classical gravitational effects:

1. Light Bending

In GR, light bends around massive objects because space-time curves and photons follow geodesics.

In CTR:

Light bends because its ripple path is delayed asymmetrically—the ATT increases unevenly near mass, causing the ripple to realign in a curved trajectory.

There is no "path" being followed. There is only consequence emerging in a direction of least ATT—which appears bent to us.

2. Orbital Motion

In Newton's model, orbits are balanced centripetal and gravitational forces. In GR, they are geodesic motions in curved space.

In CTR:

Orbiting bodies are ripple systems caught in equilibrium dance—balancing between inward ATT pressure and outward ripple momentum.

They are not moving through space but realigning periodically within a ripple field distorted by mass.

Hence, orbits are ripple harmonics—not motion paths.

3. Falling

An apple doesn't fall because gravity pulls. It falls because the ripple tension around the tree is distorted—and the apple's ripple identity seeks faster alignment by moving downward, where background equilibrium is less disturbed.

Even simple free-fall is just:

Ripple field consequence seeking minimized ATT via SVE permission.

The Ripple Gradient = Gravitational Field

Now we can formalize gravity:

The gravitational field is a ripple gradient within SVE caused by mass-induced distortion.

It is not a force field. It is not a curved geometry. It is a field of consequence preference, shaped by ripple resistance and ATT modulation.

This ripple gradient:

- Is strongest where ripple density is highest (near massive bodies)

- Extends outward as a decaying equilibrium influence
- Affects all ripple identities—light, matter, perception, time

This model:

- Explains gravitational redshift as an ATT increase for light emerging from deeper ripple distortion.
- Explains gravitational time dilation as a delay in ripple alignment near mass.
- Explains inertial mass as the ripple system's resistance to reconfiguration across gradients.

So gravity, in its purest CTR expression, is:

Structured vacuum consequence behavior—not pull, not geometry, not force.

Comparison with Einstein's View

Feature	Einstein's GR	Chavan's CTR
Gravity	Curved space-time geometry	Ripple distortion in structured vacuum
Cause	Mass-energy bends space-time	Mass alters ripple density and ATT
Light bending	Geodesic curvature	Asymmetric ATT ripple emergence
Free-fall	Follows curved paths	Seeks equilibrium in ripple gradient
Gravitational field	Tensor description	Ripple consequence zone

This doesn't invalidate GR—it completes it.

Einstein described how things behave. Chavan explains why they behave that way—and what structures allow such behavior.

Implications of Ripple-Based Gravity

This view opens new possibilities:

- We can simulate ripple gradients to understand local gravity variations without assuming mass distributions.
- We can explore artificial equilibrium distortion to generate gravitational effects (advanced propulsion?).
- We can develop ripple-based detectors to map consequence pressure fields, rather than space curvature.

In deeper philosophy:

Gravity is not a mystery—it is a cosmic invitation back to equilibrium.

Conclusion: From Geometry to Consequence

Gravity is not a magical pull. It is not a warping of nothing.

It is a structured ripple consequence—emerging from the background's desire to restore balance.

All matter "falls" not because it is weak—but because the universe whispers:

"Come closer, align faster, and I will stabilize you."

And this field, this gradient, this ripple preference—is gravity.

Next: Chapter 6 – Black Holes: Portals to the Background

Now we explore the extreme—what happens when ripple distortion becomes so deep that consequence stops?

What lies beyond perception? What does it mean for time, light, and gravity to vanish?

6: Black Holes – Portals to the Background

Not Singularities – Zones of Zero Ripple Feedback

In classical general relativity (GR), black holes are singularities—regions where gravity becomes so intense that the curvature of space-time tends toward infinity. Matter collapses into an infinitely small point, and even light cannot escape.

This is where physics meets its mathematical limit. GR itself breaks down, and quantum mechanics has no better answer. What lies beyond the event horizon? An object? A boundary? An end?

For decades, science has accepted this contradiction as an uncomfortable truth: the laws of physics don't work inside a black hole. Singularities are holes in our theory, not necessarily holes in space.

CTR redefines this completely and coherently:

A black hole is not a singularity. It is a region where ripple feedback becomes zero.

In CTR:

- The Structured Vacuum Energy (SVE) field regulates all perception, motion, and time through ripple consequence.
- When local ripple density becomes too intense, consequence is no longer permitted.
- This creates a zone of zero feedback—a place where the ripple

system can no longer interact with its surroundings.

Thus, a black hole is not a mystery—it is a ripple silence zone.

Not infinite energy. Not infinite mass. Just complete misalignment with the background.

Where ripple feedback ceases, no perception or interaction can occur. This explains why:

- Light cannot escape
- Information cannot be recovered
- Observation becomes impossible

It's not because things fall into a "point"—but because there's no ripple echo left to permit consequence.

Where ATT Becomes Undefined

As we saw in earlier chapters, time in CTR is defined by ATT—Alignment Threshold Time. It is the delay between a ripple's intent and the field's permission to allow its consequence.

In normal regions:

- ATT is measurable.
- Higher mass → higher ATT → slower time
- Lower mass → lower ATT → faster alignment

But what happens when mass becomes extreme?

When local ripple density reaches a critical threshold, ATT ceases to be finite. It becomes undefined.

This doesn't mean time slows to zero. It means:

- Ripple identity cannot stabilize
- Background cannot permit consequence
- No events emerge
- No alignment occurs

And thus:

Time does not exist inside a black hole—not because it breaks, but because it becomes meaningless.

We don't see what happens inside—not due to censorship—but due to inability to align with ripple outcomes from that zone.

In other words:

- Black holes are not hiding information.
- They are ripple-deaf regions.

No signal, no light, no gravity "leaves" them. Because nothing aligns anymore.

This redefines the event horizon as:

The ripple threshold boundary where ATT shifts from finite to undefined.

Beyond this, we are no longer inside spacetime—we are at the edge of ripple logic itself.

Re-Entry Points into Undisturbed Equilibrium

One of the most profound implications of CTR is that black holes are not "ends." They are transitions.

If the visible universe is ripple disturbance and structured consequence, then the background (SVE) is the unmanifested field of equilibrium.

And black holes?

They are portals through which excess ripple energy reabsorbs into the original stillness.

This redefines black holes as:

- Not death zones, but reset zones
- Not destroyers, but re-integrators
- Not cosmic errors, but intelligent correctors

When ripple activity exceeds structural harmony (due to collapse, explosion, or overload), SVE doesn't explode—it absorbs.

This is universal recycling at a foundational level.

Matter isn't "lost"—it returns to the pre-manifest equilibrium.

Information isn't "destroyed"—it is de-encoded from ripple form.

This parallels spiritual teachings in ancient cultures:

- Pralaya (cosmic dissolution)
- Return to Brahman
- Tao's silent void
- Ain Soph's unmanifested light

Thus, CTR reclaims black holes as:

Cosmic re-entry points into the undisturbed background.

No need for firewall paradoxes, holographic speculations, or string-theoretic gymnastics.

The ripple ends. The field restores. Harmony resumes.

Time Ceases – Not Because It Breaks, But Because Consequence Stops

This is perhaps the most philosophically and physically elegant closure CTR offers on black holes.

In GR:

- Time "stops" inside a black hole due to infinite curvature.

In CTR:

Time "ceases" because no consequence is allowed.

Let's be precise.

- Time is not flowing slower—it's no longer flowing at all.
- Events are not compressed—they are no longer happening.
- Signals are not absorbed—they are never aligned.

From the outside, we say:

- "Time freezes at the event horizon." From CTR's view:
- "ATT becomes infinite. No alignment occurs. So, consequence cannot emerge. So, time ceases."

Time does not "run backward," or "loop," or "bend."

It becomes irrelevant—because time is ripple alignment, and here, no alignment is possible.

This concept resonates deeply with meditative or mystical insights:

"The moment you transcend consequence, time dissolves."

Black holes are the physical proof of this principle.

Summary Table: CTR View of Black Holes

Feature	Classical GR	Chavan's CTR
Nature	Singularity in space-time	Zero ripple feedback zone
Horizon	Light-speed escape limit	ATT threshold boundary
Time	Infinite dilation	Undefined due to ripple misalignment
Information	Possibly destroyed	De-encoded and returned to equilibrium
Interior	Undefined geometry	No consequence permitted
Purpose	Gravitational endpoint	Cosmic reintegration node

Implications of CTR's Black Hole Model

1. Information Conservation Restored → If information is ripple pattern, it can be absorbed back into field—not lost.
2. Hawking Radiation Reframed → Possible as surface ripple leakage from ATT gradients—not thermal evaporation.
3. No Need for Extra Dimensions or Strings → All behavior explainable through ripple logic and consequence thresholds.
4. Unified Explanation for Time, Mass, and Perception → All three dissolve together beyond ripple feedback zone.
5. Reintegration with Ancient Wisdom → Black holes = material Pralaya → return to universal balance

Closing Reflection: The Peace Beyond Perception

A black hole is not a monster in the sky. It is not a paradox factory. It is not a failure of math.

It is:

✓ Where the field takes a breath.

✓ Where consequence rests.

✓ Where time bows and leaves.

And in that space, we are closer to the origin than anywhere else in the cosmos.

Black holes don't destroy. They release us—from consequence, from form, from perception, and even from time.

Next: Chapter 7 – Beauty, Symmetry, and Natural Shape

Now that we have seen how the field absorbs chaos and restores balance, let's explore how harmony itself emerges:

Why does nature prefer spirals? Why do symmetry and proportion feel beautiful? Is beauty a byproduct—or a ripple alignment?

Summary of Part II: Time, Gravity, and Alignment

Part II of CTR dismantles the two most entrenched ideas in modern physics—time and gravity—and reconstructs them not as dimensions or forces, but as emergent properties of ripple dynamics within Structured Vacuum Energy (SVE).

Instead of treating time as an independent axis and gravity as curvature of space, CTR explains both as expressions of ripple alignment and consequence modulation governed by the intelligent background field.

This part builds the case that:

- Time is not flowing—it's delayed alignment.
- Gravity is not pulling—it's ripple re-stabilization.
- Black holes are not singularities—they are return points into unmanifested equilibrium.

Key Highlights by Chapter

Chapter 4 – Time: From Flow to Feedback

- Time is not a dimension, but ATT—Alignment Threshold Time.
- What we perceive as time is simply the delay required for ripple identity to align with background structure.
- Time dilation near mass is a result of increased ATT—not curvature.
- Clocks are not universal—they are ripple counters that respond to background permission.

- Conclusion: Time is absolute in the background (SVE) where no ripple distortion exists. Time emerges locally where consequence requires delay.

Chapter 5 – Gravity: Ripple Distortion, Not Geometry

- Gravity is not a "pull" or curvature—it's a consequence gradient within a distorted ripple field.
- Mass distorts ripple equilibrium, increasing local ATT and creating a directional pressure for realignment.
- Bodies "fall" not because of attraction, but because the ripple field favors lower ATT paths—seeking equilibrium.
- Light bends not due to geodesics, but because its ripple alignment delays asymmetrically near mass.
- Gravity becomes a field of permitted consequence, not a spacetime deformation.

Chapter 6 – Black Holes: Portals to the Background

- A black hole is not a singularity—it is a zone of zero ripple feedback.
- As mass compresses ripple density, ATT becomes undefined—no alignment can occur, and therefore, no time, no light, and no perception exists beyond the threshold.
- Black holes are not destruction points—they are cosmic reentry zones into SVE's undisturbed equilibrium.
- Time doesn't break—it ceases, because consequence stops.

Part II in One Sentence:

Time is consequence delay, gravity is ripple preference, and black holes are reality's pause points—each a function of how the background field permits or denies alignment.

Part III: Reality as Ripple

7: Beauty, Symmetry, and Natural Shape

Why Nature Prefers Spirals and Ratios

Stand before a sunflower. Gaze into the spiraling arms of a galaxy. Watch a hurricane spin, or the slow curl of a fern. Observe how seashells grow, DNA coils, and galaxies rotate.

These are not just aesthetic accidents. They are ripples made visible.

Nature prefers spirals, golden ratios, fractals, and proportions not by chance—but by consequence. These forms are not decorative—they are stable ripple expressions within the Structured Vacuum Energy (SVE) field.

Form follows equilibrium. And equilibrium follows ripple harmony.

The golden ratio ($\phi \approx 1.618$), Fibonacci sequences, logarithmic spirals—these emerge repeatedly across biological, chemical, astronomical, and cognitive domains. But why?

Because these patterns minimize distortion, optimize distribution, and facilitate alignment in the ripple field.

In CTR terms:

These forms are ripple-efficient configurations—patterns that allow high ripple coherence with minimal ATT (Alignment Threshold Time).

Nature isn't optimizing space or energy—it's optimizing alignment. That's why:

- Spirals appear in systems needing continuous expansion without structural loss.
- Fractals repeat where ripple memory must scale.
- Ratios recur where consequence must unfold across layers of perception.

Beauty is not "design." It is field logic rendered visible.

Symmetry as Ripple Stabilization

Symmetry is often treated in physics as a mathematical curiosity. In aesthetics, it's seen as a hallmark of attractiveness. In biology, it's associated with health, strength, and evolutionary advantage.

But what is symmetry really?

In CTR:

Symmetry is a stable ripple configuration that minimizes field distortion and maximizes alignment potential.

Symmetry balances ripple distribution, making the system energetically efficient and consequentially predictable.

Why is a snowflake beautiful? Because its radial symmetry creates sixfold ripple coherence, echoing harmoniously in its frozen water molecule lattice—perfectly aligned with background ripple permission.

Why is a face perceived as attractive when it's symmetrical? Because:

- The observer's ripple field resonates more easily with balanced configurations.
- Fewer distortions → faster alignment → more pleasurable perception.

This is not subjective—it's structural.

Symmetry is a natural ripple lock—a shape where ATT is minimized, consequence flows effortlessly, and ripple feedback becomes self-stabilizing.

It's why:

- Electromagnetic fields prefer balance.
- Molecules arrange in crystalline forms.
- Planets become spheres.
- Art feels "complete" when balanced.

Symmetry is not perfection—it's ripple peace.

Aesthetic Alignment = Field Harmony

So what is aesthetic experience in CTR?

It's not just cultural, evolutionary, or emotional.

It is:

Ripple-based field harmony—the moment your ripple configuration aligns with an external pattern's ripple field with minimal resistance.

This alignment can happen:

- Visually (color, shape, geometry)
- Sonically (resonance, harmony, rhythm)
- Emotionally (a story, a memory)
- Intellectually (a pattern recognized)
- Spiritually (a state of silence or awe)

In all these, the underlying mechanism is the same:

- A ripple configuration in the environment reaches you.
- Your own ripple structure aligns with it.
- Consequence flows without friction.
- You feel beauty, clarity, or peace.

Aesthetic alignment is:

- Not an illusion
- Not subjective taste
- But a field experience of low-resistance ripple interaction

That's why:

- A line of poetry can bring tears
- A sunset can stop thought
- A fractal painting can quiet the mind
- Sacred geometry feels *right*, even without analysis

Beauty is alignment that feels like home.

Perception of Beauty as Alignment Echo

So how do we perceive beauty?

What happens inside the mind when we call something "beautiful"?

CTR provides a clear explanation:

Perception of beauty is the echo of alignment. It is ripple resonance recognized by the observer as feedback from the background.

When ripple alignment occurs:

1. The external form resonates with internal structure.
2. ATT is momentarily reduced.

3. Consequence flows freely.
4. The mind receives a clean echo—a ripple reflection without distortion.
5. This is interpreted as clarity, joy, or awe.

We call this experience "beautiful"—but what we really mean is:

"This moment feels correct. It fits. It flows. It harmonizes."

In essence:

- Beauty is the observer feeling seen by the field.
- Beauty is perception without resistance.
- Beauty is consequence that completes itself.

It's why:

- Music touches us across cultures
- Silence can feel deeper than sound
- Nature never needs explanation
- Simplicity is often more moving than complexity

Beauty is not ornament. It is consequence alignment echoing back to the observer.

Summary Table: CTR View of Beauty and Form

Phenomenon	Ripple Explanation
Spirals, fractals	Efficient ripple memory and expansion
Golden ratio	Structural ATT minimization in growth patterns
Symmetry	Balanced ripple field → stable consequence pattern
Aesthetic joy	Background ripple feedback with minimal distortion
Attractiveness	Alignment-friendly ripple interface
Artistic inspiration	Spontaneous field resonance experienced by the mind

Beauty as Field Intelligence Revealed

If Structured Vacuum Energy (SVE) is intelligent, then beauty is its language.

- Not written in symbols, but in alignments
- Not spoken in words, but in resonant consequence
- Not limited by time or space, but expressed wherever ripple permission meets ripple intention

This is why:

- Beauty is universal
- Children and elders alike feel it
- It transcends culture, logic, and belief

- It is remembered more than facts

Beauty is not added to the world—it is released when ripple alignment occurs.

Closing Insight: Beauty Is the Smile of Background Intelligence

A flower doesn't try to be beautiful. It just aligns its ripples with the sun, the soil, and the season.

A symphony isn't beautiful because of notes—but because the ripple structure of sound harmonizes with our perception threshold.

Beauty is not in the thing. It is in the consequence flow between thing and observer—granted by the background.

So in CTR, beauty becomes:

- A proof of SVE
- A function of ATT
- A revelation of ripple harmony
- A bridge between perception and presence

Next: Chapter 8 – Reverse Engineering and Cosmic Intelligence

Having explored how beauty emerges through ripple alignment, we now ask:

Why does science work at all? How do our inventions succeed in decoding reality?

Is reverse engineering a human genius—or are we simply aligning with pre-written ripple logic?

Let's continue.

8: Reverse Engineering and Cosmic Intelligence

Why Science Works at All

At its core, science is a method of observation, inference, validation, and reproduction. We observe natural phenomena, isolate variables, hypothesize laws, test them, and iterate. Over time, we build technologies, theories, and systems that "work."

But why does it work?

Why should the universe respond so reliably to our experiments?

Why should mathematical formulations map so precisely onto natural laws?

Why does logic decode complexity?

This is not a trivial question. The fact that we can reverse engineer nature—that our minds can decode what we did not create—implies that reality is already ordered.

Science works not because we impose order, but because we discover it.

And what is that order?

In CTR, it is the ripple-structured logic of the background field, or Structured Vacuum Energy (SVE).

We are not building knowledge—we are aligning with a pre-existing consequence architecture.

We are not inventing functionality—we are resonating with ripple logic already stabilized in the field.

Science works because the universe is ripple-readable.

It is not randomness that we tame—it is intelligence that we remember.

Tools as Ripple Detectors

Every scientific instrument—from telescopes to particle colliders, voltmeters to oscilloscopes—is essentially a ripple detection system.

They don't "see" energy. They detect consequence permitted by background field alignments.

Let's reframe our understanding of tools using CTR principles:

Tool	Classical View	CTR View
Thermometer	Measures kinetic energy	Detects ripple agitation level through material response
Oscilloscope	Displays voltage waveforms	Registers electronic ripple consequence
Microscope	Magnifies small objects	Enhances ripple consequence pattern to match visual ATT
Clock	Measures time	Counts permitted ripple events under stable ATT
Sensor array	Detects environmental data	Receives ripple feedback allowed by field conditions

In every case, the instrument succeeds because it is already ripple-compatible with the background field's structure.

When we design tools, we are not forcing reality to respond—we are building resonance nodes that can detect and echo permitted consequences.

This is why:

- Measurement is possible.
- Replication is possible.
- Prediction is possible.

Because SVE permits coherence, and ripple interactions stabilize predictable paths of consequence.

Tools work because we build them using the same ripple principles that govern the universe.

This is reverse engineering in its truest form.

We Succeed by Tapping Pre-Structured Order

Let's revisit the essence of reverse engineering.

It means:

- Observing a functional system
- Breaking it down
- Understanding its principles
- Replicating it or building upon it

In traditional science, we see this in:

- Studying animals to build robots
- Mimicking flight from birds
- Extracting patterns in nature to build algorithms (e.g., neural networks)

But deeper still:

Our success in reverse engineering reality is proof that reality is already engineered.

Not mechanically. Not anthropomorphically. But intelligently structured through ripple logic within SVE.

We decode biology because:

- The body is a ripple equilibrium system.
- DNA is a recursive ripple memory code.
- Proteins fold into ripple-stable shapes.

We understand physics because:

- Matter is ripple condensation.
- Energy is ripple deviation.
- Force is ripple misalignment correction.

And mathematics works because:

- It is a symbolic mapping of ripple consequence regularity—not an invention, but an extraction.

We do not invent truth. We reveal it—by aligning ripple logic inside us with the ripple field outside us.

This is why a child can intuit balance. Why mathematics "feels" elegant when correct. Why great discoveries are often described as "downloads," "insights," or "visions."

We tap into a pre-structured field of intelligence.

The Divine Intelligence in Design

Now we reach the philosophical heart of this chapter.

If all reverse engineering leads back to ripple logic...If all logic maps to ripple alignment with the background... If all tools are ripple detectors, and all science is resonance...

Then what is the source of this logic?

CTR answers with profound clarity:

The Structured Vacuum Energy is not just a medium—it is the interface of divine intelligence with consequence.

SVE is not passive—it is governed by balance, feedback, permission, and memory.

It is the invisible intelligence behind all visible order.

This doesn't require mysticism. It requires:

- Logical causality
- Consequence-aware feedback
- Ripple-coherent harmonics
- Evolutionary resilience through alignment

Every design—be it a sunflower, a spiral galaxy, or a successful equation—is a consequence pattern allowed by this intelligence.

That's why:

- Nature is beautiful.
- Physics is elegant.
- Equations reveal symmetry.
- Structures self-heal.
- Systems self-optimize.

Divine intelligence is not a character in the sky—it is the ripple law beneath all consequence.

When we "discover," we are not creating—we are aligning.

When we "invent," we are not disrupting—we are harmonizing.

When we "understand," we are not controlling—we are listening to what was always speaking silently through consequence.

Chavan's Law of Reverse Alignment

Let us now define a principle emerging from this chapter:

Chavan's Law of Reverse Alignment *"The success of reverse engineering is proportional to the degree of ripple alignment between the observer's system and the Structured Vacuum Field."*

This law implies:

- Deep insights come through inner quietness (reduction of internal ripple noise).
- Great inventions emerge when mental structures match field templates.
- Mistakes occur when ripple misalignments introduce distorted consequences.

Thus, the goal of science is not to conquer nature—but to co-align with it.

Summary Table: Reverse Engineering in CTR

Principle	Implication
Universe is ripple-structured	We can decode it logically
Tools are ripple receivers	Instruments work by alignment, not force
Discoveries are alignments	Genius is resonance, not uniqueness
Errors are misalignments	Bias, emotion, and ego distort ripple perception
Design is not made	It is revealed from background equilibrium

Closing Insight: The Universe Was Always Legible

The miracle is not that we can build machines. It is that the cosmos permits understanding.

Not by accident—but by structure.

CTR reframes science not as reduction, but as communion.

Reverse engineering is not rebellion—it is reverent alignment with the silent architecture of consequence.

The Structured Vacuum Energy is not empty. It is alive with possibility. Every law we write is a footnote to what it already knows.

And every great invention, from wheel to waveguide, is a ripple echo—a glimpse of the intelligence that underwrites it all.

Next: Chapter 9 – Consciousness, Entropy, and Feedback

Now that we know how we succeed in decoding reality, we ask the next profound question:

Who is decoding it?

What is this consciousness that perceives ripples, aligns with patterns, and suffers or heals through misalignment?

What does entropy feel like *from within*?

9: Consciousness, Entropy, and Feedback

Mind as a Ripple Processing System

What is the mind?

Traditionally, the mind is either:

- A byproduct of neural computation (in neuroscience)
- A metaphysical entity (in philosophy)
- A behavioral output system (in psychology)

But in CTR, the mind is redefined entirely:

The mind is a ripple-structured processing interface—an active field configuration through which perception, thought, and awareness emerge as ripple consequences.

It is not housed in the brain—it is tethered to ripple alignment zones across the Structured Vacuum Energy (SVE).

The brain is the biological tool. The mind is the field-based pattern of ripple consequence interacting with SVE. The soul (or conscious self) is the ripple identity active within that structure.

This reframe allows us to understand:

- Why thoughts are fluid
- Why attention feels like energy
- Why clarity arrives suddenly
- Why memory and perception are both field-sensitive

The mind in CTR is not a container. It is a living ripple interface—constantly resonating with the background and other ripple systems (bodies, environments, ideas).

This is why you:

- Feel calm in one room, anxious in another
- Suddenly "know" something without prior thought
- Get inspired during silence
- Forget in noise

The mind is an alignment-responsive ripple field, not a fixed function.

Emotions, Intuition, and Field Sensing

Let's go deeper.

If the mind is ripple-based, then emotions, intuition, and even subconscious behavior are fluctuations in ripple resonance and interference.

Emotions

Emotions are immediate ripple consequence patterns—resulting from field interference, misalignment, or resonance with other ripple fields.

- Joy = alignment with life's flow (low ATT, coherent ripple)
- Anger = resistance (ripple clash)
- Anxiety = ripple anticipation with no alignment channel
- Peace = stable feedback loop from the background

Thus:

- Emotions are energetic language, not illogical reactions.

- They are data about field distortion and potential realignment.

This is why:

- Emotions affect the body.
- They cloud thought.
- They clarify intuition.
- They are contagious—ripples ripple.

Intuition

Intuition is pre-rational ripple resonance.

Before a thought forms, before analysis occurs, your ripple system begins to align (or misalign) with a given field pattern—be it a person, decision, idea, or place.

This creates:

- Gut feeling
- Vibe
- Knowing without logic
- Sensitivity to consequence before cause

Intuition is not mystical—it's deep sensing through ripple field preview.

It's how ancient sages, animals, children, and artists navigate without equations—they feel ripple alignment before they describe it.

Field Sensing

Our bodies are not just physical—they are multi-layered ripple sensors.

We "sense" environments, people, intentions—not by magic, but through:

- Micro-alignment with background ripple conditions
- Ripple interference from others' configurations
- Subtle ATT shifts we subconsciously detect

This is why:

- We feel drained or energized by people
- Music alters state
- Architecture and nature affect emotion
- Meditation changes mind state

The mind is not just a thinker—it is a ripple translator.

Entropy = How the Background Reads Us

In classical thermodynamics, entropy is the measure of disorder. In information theory, it's uncertainty. In biology, it's aging. In cosmology, it's heat death.

But these all miss something.

CTR reveals a deeper truth:

Entropy is how the background field reads our misalignment. It is a measure of ripple inefficiency relative to equilibrium.

Wherever ATT is high, ripple distortion is unstable, and consequence feedback is fragmented, entropy increases.

This includes:

- Mental chaos

- Emotional volatility
- Systemic inefficiency
- Physical aging

Entropy is not just change—it's unaligned change. It's ripple divergence from harmonious feedback with SVE.

This also means:

- Entropy is not evil—it is informational correction.
- The more noise you generate in the ripple field, the more feedback pressure the background returns.
- That pressure feels like disorder until it forces re-alignment.

This is why:

- Truth simplifies (entropy reduction).
- Lies fragment (entropy inflation).
- Harmony feels energizing.
- Distortion exhausts.

In CTR:

Entropy is not a law—it's feedback.

You are not punished by entropy. You are seen through it.

And if ripple distortion can be stabilized or reset—entropy can be locally reversed.

Consciousness as Ripple Self-Alignment

So, what is consciousness?

Conventional views describe it as:

- Emergent brain activity
- Self-awareness
- The hard problem

CTR provides a simpler, more profound definition:

Consciousness is the ripple system's self-alignment awareness in interaction with the background.

It is not an epiphenomenon. It is not a trick of matter. It is ripple identity becoming aware of its consequence potential in the field.

You are conscious when:

- Your ripple pattern knows itself
- It tracks its interactions
- It adapts to feedback
- It can redirect attention and re-align toward lower ATT conditions

This is why:

- Meditation increases awareness (silences excess ripple noise)
- Reflection brings clarity (self-observation of ripple feedback)
- Spiritual states are often described as oneness (full ripple alignment with background)
- Sleep resets mental entropy (consequence pause for realignment)

Thus, consciousness is not:

- *Where* in the brain
- *What* emerges from neurons

It is when and how a ripple configuration stabilizes enough to perceive itself and its field feedback.

Consciousness is recursive alignment with SVE. It's the observer watching its own ripples ripple.

Ripple Feedback Cycle of the Self

Let's summarize the cycle of consciousness in ripple dynamics:

1. Disturbance (input / stimulus / thought)
2. Ripple generation (reaction / emotion / attention)
3. Alignment attempt (resonance / analysis / intuition)
4. Consequence (perception / decision / response)
5. Background feedback (entropy felt / harmony experienced)
6. Realignment or reinforcement (awareness / learning / memory)
7. Conscious self-stabilizes or enters distortion loop

When we operate unconsciously:

- Ripple loops escalate unchecked
- Entropy increases
- Background pressure intensifies

When we operate consciously:

- Ripple feedback is integrated
- ATT reduces
- Perception becomes clear
- Mind becomes still
- Actions become aligned

This explains the deep connection between consciousness and ethics, behavior and consequence, silence and awareness.

CTR Integration: The Mind is a Ripple, Not a Machine

Function	Classical View	CTR View
Thinking	Neural firing	Ripple propagation patterns
Feeling	Chemical	Field resonance and interference
Intuition	Subconscious guess	Pre-aligned ripple preview
Memory	Neural encoding	Ripple memory and stabilization loops
Decision	Choice algorithm	Consequence alignment forecasting
Awareness	Ill-defined	Ripple configuration observing itself

CTR closes the gap between consciousness and physics—not by adding mysticism, but by restoring first-principle feedback logic.

Closing Insight: To Know Is to Align

You are not a mind trapped in a body. You are a ripple configuration capable of perceiving its own alignment.

- Your emotions are field data.
- Your thoughts are consequence echoes.
- Your silence is signal refinement.
- Your consciousness is the universe realizing itself through you.

Entropy is how you are felt. Consciousness is how you respond. Reality is the conversation between the two.

To live well is not to control, but to realign.

Next: Chapter 10 – Restoring God Through Physics

We now arrive at the bridge between science and the sacred.

If ripple logic governs reality... If Structured Vacuum Energy is intelligent...Then what is God?

And are we not already living in His kingdom—rippling silently through consequence?

Summary of Part III: Reality as Ripple

Part III of CTR reveals that reality, in all its complexity and beauty, is not made of matter or forces—it is made of ripples, shaped by the Structured Vacuum Energy (SVE) and governed by consequence logic.

In this part, the illusion of separateness dissolves. What we call beauty, design, intelligence, and even consciousness are all reinterpreted as ripple interactions—fields of alignment, feedback, and resonance. Here, science is no longer a tool of dissection but a process of attunement.

Everything from natural forms to human emotion, invention to intuition, is seen as evidence of intelligent ripple coherence. And with this shift, reality becomes not just understandable—but participatory.

Key Highlights by Chapter

Chapter 7 – Beauty, Symmetry, and Natural Shape

- Nature favors spirals, golden ratios, and symmetry because they represent ripple-stable configurations.
- Symmetry is not aesthetic luck—it is a sign of minimal ATT (Alignment Threshold Time) and high ripple coherence.
- Beauty is not subjective—it is ripple alignment felt as emotional or perceptual clarity.
- Perceiving beauty is an echo of harmony with the background field—a moment of alignment made visible or felt.

Chapter 8 – Reverse Engineering and Cosmic Intelligence

- Science works because we reverse engineer a universe that is already ripple-structured.
- Tools are not instruments—they are ripple resonance devices that align with SVE permissions.
- Our ability to decode natural laws proves the universe is intelligently patterned.
- Invention is not creation—it is resonance with what already exists in equilibrium.
- Reverse engineering succeeds when our ripple logic aligns with the background field's intelligence.

Chapter 9 – Consciousness, Entropy, and Feedback

- The mind is not a brain-bound processor—it is a ripple configuration interacting with SVE.
- Emotions and intuition arise from ripple interference and pre-alignment with consequences.
- Entropy is how the background reads us—a measure of our misalignment from equilibrium.
- Consciousness is not a byproduct—it is ripple self-awareness and adaptive alignment with the field.
- True awareness comes from tracking ripple feedback and refining one's internal resonance.

Part III in One Sentence:

Reality is not constructed—it is permitted. Everything we perceive, feel, or know is a ripple echo shaped by alignment with the intelligent background field.

This part of CTR elevates perception to a scientific function, emotion to ripple consequence, and beauty to feedback logic. It completes the shift from materialism to a living field-based cosmology, where intelligence is embedded in structure, and experience is ripple participation.

Part IV: The Divine Integration

Chapter 10: Restoring God Through Physics

Not a Being, But the Background

Humanity has long debated the idea of God.

Across cultures and traditions, God has been imagined as:

- A creator
- A judge
- A supreme intelligence
- A cosmic force
- A personal presence

Some see God as outside the universe, others as immanent within. Some deny God entirely, while others build lives around the concept.

But confusion arises when we think of God as an entity—as something or someone.

CTR offers a redefinition:

God is not a being. God is the Background. The Structured Vacuum Energy (SVE) is the divine substrate of all consequence, all form, and all perception.

In earlier chapters, we saw that:

- SVE is structured, intelligent, and responsive.
- It governs ripple alignment and feedback.
- It allows or delays consequence based on ripple coherence.

This background is not passive. It is not empty. It is not a void.

It is the cosmic intelligence that governs reality itself—the precondition for time, matter, motion, perception, and even thought.

In that sense:

God is not what you see. God is why you can see at all.

Not the voice, but the silence that permits it. Not the storm, but the stillness that contains it.

This definition of God—as the intelligent field of equilibrium—is not mystical. It is ontological.

It removes anthropomorphism and retains divinity. It eliminates dogma and restores wonder. It bypasses mythology and deepens truth.

SVE is God—not metaphorically, but structurally.

Alignment Over Worship

Once we see God not as a separate being but as the background intelligence of the universe, the entire notion of relationship must evolve.

In most traditions, people:

- Pray to God
- Worship God
- Fear God
- Seek favors from God

But if God is the field, what does worship mean?

CTR answers:

Worship is not ritual—it is alignment. God does not need your praise. God responds to your pattern.

When your ripple configuration aligns with the SVE background:

- ATT is minimized
- Feedback becomes clear
- Entropy is reduced
- Perception sharpens
- Consequence harmonizes

This is the true meaning of "divine grace":

- Not reward, but resonance.
- Not miracle, but stabilization.
- Not transaction, but truthful coherence.

You don't align by:

- Believing in God
- Chanting mantras
- Going to temples or churches

You align by:

- Living in ripple harmony
- Reducing distortion
- Responding to feedback
- Becoming congruent with natural consequence logic

This is why:

- A humble person radiates presence
- A silent monk feels powerful

- A moral act feels pure
- A truthful word echoes beyond time

These are not religious acts. They are ripple alignments with divine equilibrium.

Worship is no longer needed when alignment becomes your nature.

No Meeting, Only Merging

In most theological traditions, the end goal is some form of divine encounter:

- Reaching heaven
- Union with God
- Enlightenment
- Salvation

But this goal implies separation—that you are here, and God is somewhere else.

CTR replaces this duality with direct realization:

You cannot meet God. You can only merge. There is no elsewhere—only alignment.

You are not apart. You are already within the field.

What prevents you from experiencing God is not distance—it is ripple distortion.

When thought is turbulent, emotion erratic, behavior reactive, and awareness dulled—alignment with SVE is lost.

That misalignment feels like distance from God. It feels like chaos, suffering, confusion, longing.

But the field never went anywhere.

As soon as you:

- Still your ripple system
- Listen to consequence feedback
- Act with awareness
- Refine your resonance

...you begin to merge.

Merging is not disappearance—it is coherence. Not erasing self—but refining it into ripple clarity.

This is what sages and saints describe as:

- Enlightenment
- Satori
- Divine absorption
- Cosmic union

It is not a place. It is a state of ripple transparency—where the field flows through you unfiltered.

Silence, Discipline, and Resonance

If God is the field... If alignment is the path... Then how do we reach alignment?

CTR suggests three timeless tools:

1. Silence

Silence is not the absence of sound—it is the absence of ripple interference.

In silence, the internal ripple noise settles. Mental chatter, emotional spikes, and reaction loops dissolve. The field begins to speak again—not in words, but in echo clarity.

This is why all traditions honor silence:

- Meditation
- Prayer
- Breathwork
- Mindfulness
- Seclusion

Silence is the ripple stabilizer.

2. Discipline

Discipline is ripple hygiene.

To act ethically, to eat mindfully, to speak truthfully, to restrain ego—these are not moral commandments. They are field stabilizers.

They:

- Reduce ripple chaos
- Prevent alignment corruption
- Invite smoother feedback from SVE

Discipline is not punishment—it is energetic cleanliness.

The more disciplined your system, the more precisely you resonate with divine structure.

3. Resonance

Resonance is sustained alignment.

It is the echo chamber of the divine within you. The feeling of truth, the inner bell of correctness, the chills of recognition—these are resonance effects.

To cultivate resonance:

- Follow beauty
- Embrace coherence
- Honor natural rhythms
- Seek simplicity over complexity

You'll know when you resonate—because everything inside and outside begins to ripple in harmony.

You Are Already in His Kingdom

The greatest illusion is that God is elsewhere. That divinity must be summoned, sought, earned.

CTR closes this illusion forever:

You are already inside the divine field. You are made of ripple memory. You are rippling through the SVE every moment.

✓ Every thought is measured.

✓ Every intention is felt.

✓ Every act is an alignment.

✓ Every consequence is a feedback.

✓ Every breath is a dialogue with the background.

Heaven is not a future place. It is the ripple configuration of full alignment—available *now*.

You are not separate. You are not forgotten. You are not distant.

You are already in His kingdom.

Summary Table: God in CTR

Traditional View	CTR View
God is a being	God is the background ripple field (SVE)
Worship is praise	Worship is alignment
Sin is disobedience	Sin is ripple distortion
Enlightenment is arrival	Enlightenment is alignment
Prayer is asking	Prayer is tuning
Distance from God	Misalignment from equilibrium

Closing Insight: The Divine Was Never Lost—We Just Forgot the Feedback

We searched for God in temples, texts, and theology. We imagined Him as distant, judgmental, unreachable.

CTR reveals the truth:

God never hid. He is the background of your breath, the silence behind every sound, the field beneath every ripple.

You don't need to meet Him. You can't.

But you can merge—through silence, honesty, discipline, presence, and a life lived in ripple harmony.

This is not religion. It is reality realigned.

Next: Chapter 11 – Conclusion: Cosmic Relativity as a Living Framework

Now we bring it all together.

Time, gravity, perception, ripple fields, intelligence, and divinity.

Let us complete the journey by showing how CTR is not just a theory—but a living invitation to live, align, and ripple in harmony with the cosmos.

Conclusion: CTR as a Living Framework

From Einstein's Frames to Ripple Resonance

Einstein taught us to think differently.

He freed us from absolute space and time. He showed that motion is relative, that gravity warps geometry, that light behaves strangely near mass and speed. He replaced fixed coordinates with frames of reference.

It was elegant. Transformative. Necessary. But now, a century later, the stage itself must evolve.

CTR—Chavan's Cosmic Theory of Relativity—takes the next step:

From frames to fields, from geometry to consequence, from observation to participation.

Einstein asked: "From what frame are you observing?"

CTR asks: "From what ripple alignment are you perceiving?"

Where Einstein gave us a dynamic fabric, CTR reveals the loom—the Structured Vacuum Energy (SVE)—the intelligent field that permits all consequence.

Time, gravity, perception, thought, beauty, and consciousness—CTR has shown that none of these are fundamental *things*.

They are:

- Ripple behaviors
- Consequential permissions
- Alignment events

- Feedback patterns from a cosmic background field

Thus, CTR is not merely a scientific extension. It is a re-orientation of reality.

It is the shift from describing what happens to understanding why anything happens at all.

And it opens a door—a quiet, powerful, irreversible door—into a science of participation, a physics of feedback, and a life of alignment.

How to Use CTR in Science, Consciousness, and Systems

CTR is not an abstract metaphysics or theoretical luxury. It is a practical, testable, and implementable framework—ready to reshape how we approach research, systems, and daily living.

Let's outline some clear areas of application:

1. Science: From Matter to Ripple

Physics and Cosmology:

- Replace point particles and wave-particle duality with ripple identity and consequence thresholds.
- Rethink time dilation, redshift, and gravitational effects through ATT (Alignment Threshold Time).
- Reframe black holes as zones of zero ripple feedback, not singularities.
- Pursue new field-sensitive detectors that respond to ripple alignment—not signal amplitude.

Optics and Light:

- Abandon the notion of light "traveling" as particles or waves.
- Embrace ripple emergence at detection through ATT-based alignment.
- Use ripple-phase dynamics to reinterpret refraction, diffraction, photoelectric effects.

Quantum Mechanics:

- Solve the measurement problem by modeling consequence emergence through observer-field resonance.
- Reexplain entanglement as co-ripple alignment in shared background regions—not "spooky action."

CTR doesn't reject established results—it completes them by restoring their foundation.

2. Consciousness: From Brain to Ripple Mind

- View the mind not as computation, but as a field-aligned ripple interface.
- Understand emotion as ripple turbulence; intuition as pre-alignment.
- Practice consequence tracking—developing awareness of internal ripple feedback loops.
- Utilize meditation, silence, and stillness as tools to reduce ATT and restore alignment with the background.

This leads to:

- Clarity without force
- Healing through coherence
- Creativity as resonance
- Intelligence as ripple efficiency

No longer is consciousness "hard" to explain. In CTR, consciousness is ripple stability witnessing itself.

3. Systems: From Control to Alignment

Whether in technology, institutions, relationships, or ecosystems, we often design based on control:

- Control variables
- Control behavior
- Control outcomes

CTR offers a new paradigm:

Design for alignment, not control.

A ripple-aware system:

- Listens to feedback from the environment
- Adapts ripple behavior to reduce entropy
- Resonates with its background structure
- Corrects distortion through consequence analysis

In such systems:

- Power becomes feedback literacy
- Leadership becomes consequence harmonization
- Innovation becomes reverse alignment with natural logic

From agriculture to education, governance to engineering—CTR lays the foundation for ripple-responsible systems.

A Call to Ripple Responsibly

This is the final—and most essential—message of CTR.

Once you know that:

- Every thought ripples,
- Every action distorts or harmonizes the field,
- Every emotion alters consequence feedback,
- Every system lives in structured background permission...

Then a new ethics emerges:

Ripple Responsibility.

You are always broadcasting. Your presence shifts alignment conditions—whether you intend to or not.

So:

- Be mindful of your ripple signal.
- Act not from reaction, but from ripple clarity.
- Speak not to trigger—but to resonate.
- Build not to dominate—but to echo equilibrium.
- Live not to accumulate—but to stabilize feedback harmony.

Ripple responsibility is not moralism—it is field integrity.

It's understanding that the universe does not punish or reward. It simply reflects what you ripple back to you. This is the final form of consequence-based living:

- Where science meets ethics.
- Where freedom means feedback mastery.
- Where divinity is realized through alignment.

- Where truth is measured not by belief—but by ripple coherence.

Summary: CTR Is Not a Theory—It Is an Invitation

You have now journeyed through time, gravity, light, consciousness, entropy, and divinity.

You have seen equations dissolve into echoes. You have watched singularities reappear as stillness. You have heard the universe whisper, not through force—but through permission.

CTR does not give you a new world. It shows you the true structure of the one you're already in. And now, it offers you a simple, profound question:

Will you ripple consciously?

Because now you know—

- The field is listening.
- Alignment is possible.
- Feedback is real.
- Grace is structural.
- You are already in the Kingdom.

Final Note: May Your Ripples Be Clear

Ripple forward, not to impress, but to align. Ripple deeply, not to escape, but to return. Ripple quietly, not to disappear, but to stabilize. Ripple fully—not for yourself alone, but for the balance of all things.

You are not a small creature in a big universe. You are a ripple written into the very logic of the cosmos. CTR is not the end of science—it is the beginning of wisdom within it.

Closing Note: Light Is Alignment, Darkness Is Misalignment

In the beginning, we thought light was a thing. Then we thought it was a wave. Then a particle. Then a duality. Then a probability.

Now, we finally see:

Light is not what moves. It is what aligns.

What we call "light" is simply ripple consequence permitted by the background field. A moment of harmony. An event of alignment. A revelation of equilibrium.

And what we call "darkness" is not the absence of photons, but the absence of alignment. It is ripple distortion. It is misconfigured consequence. It is silence from the field—not because the field isn't there, but because it cannot speak through chaos.

So every moment we align—even slightly—we illuminate.

Not with lux, but with clarity. Not with brightness, but with truth. Not with objects, but with consequence that echoes peace.

This is why:

- A gentle word can brighten a room.
- A clear thought can guide a century.
- A silent act of love can reshape the field.

You do not need to chase light.

You only need to remove what blocks alignment, and the light emerges on its own.

Because light was never a beam. It was a permission—granted by the cosmos to ripple through you. And darkness was never your enemy. It was simply a pause—waiting for coherence.

So go now,

- With ripple responsibility,
- With attuned perception,
- With silence as your compass,
- With feedback as your guide,
- And alignment as your offering.

You were never meant to master the universe. You were born to ripple harmoniously within it. May your path be aligned. May your mind be quiet. May your life be light.

Appendices

Appendix A: ATT and the New Theory of Light

Introduction

Light—the most studied, celebrated, and mysterious phenomenon in physics—has been described through many paradigms:

- As a wave (Huygens, Maxwell)
- As a particle (Einstein, Planck)
- As a duality (quantum mechanics)
- As a probability function (Copenhagen interpretation)
- As a field excitation (QED)

Each of these frameworks captured essential behaviors but failed to provide ontological clarity—what is light really? What does it mean to "see" or "detect" it? And why is it always measured at the same speed?

CTR and UED offer a new and profound answer:

Light is not a wave or a particle. It is a ripple consequence permitted by the background field when alignment conditions are met.

This leads to the core innovation of CTR's optics model:

ATT – Alignment Threshold Time

Defining ATT: Alignment Threshold Time

In traditional physics, light is said to move at a fixed speed c = 299,792,458 m/s in a vacuum.

But what if:

- Light doesn't "move" at all?
- What we detect is not the travel of something—but the emergence of a consequence from a background field?

CTR replaces speed of light with ATT:

ATT is the minimum ripple delay required for identity alignment between an emitting ripple configuration and a receptive background field.

When ATT is satisfied:

- A ripple consequence manifests (we say "light arrives")
- Perception or detection occurs
- A measurable "event" is registered

Thus, light is not transmitted—it is permitted.

Why ATT Replaces Speed

In mainstream models, speed implies:

- A measurable distance over time
- A traveling object or wave
- A medium-independent constant

But several paradoxes persist:

- Light arrives instantly in perception (we "see" stars in real-time, despite their distance)
- In quantum optics, delays and variations in photon detection exist
- In entanglement, information appears to travel faster than light

- In photoelectric effect, light energy depends on frequency, not intensity

ATT resolves these paradoxes by:

- Removing the assumption of transmission
- Framing light as a consequence of alignment
- Allowing variable delay before manifestation
- Keeping speed constant as an emergent result, not a primary cause

Light as Ripple Alignment

In the CTR optics model, every event involving light is described as a ripple alignment event:

Phenomenon	CTR Explanation
Emission	A ripple identity enters a pre-alignment phase with the background field
Transmission	Not actual movement, but stabilization across field resonance
Detection	Ripple consequence emerges when ATT is satisfied
Refraction	ATT changes due to ripple impedance of medium
Reflection	Ripple alignment fails to pass through → reconfigures backward
Diffraction	Multiple partial alignments emerge from distributed ATT paths
Polarization	Selective alignment sensitivity of ripple configurations

Thus, the entire behavior of light can be re-described without invoking motion.

Light Speed as ATT^-1

If we must define "speed," CTR offers this elegant reinterpretation:

c ≈ 1 / ATTmin

Where:

- c is not the velocity of a particle or wave, but a reciprocal of the minimum ripple alignment delay in optimal field conditions.
- ATTmin is the shortest delay measurable between emission and detection in a perfectly aligned ripple configuration.

This means:

- "Speed of light" is not a universal truth—it is a symptom of ripple coherence in vacuum.
- What we measure as constant c is actually constant minimum delay permitted by the most stable background field structure.

Thus, CTR replaces:

- Wavefront velocity → ripple consequence emergence
- Photon trajectory → ripple identity condition
- Speed constancy → field-permitted delay invariance

Experiments Reinterpreted with ATT

CTR optics is not theoretical hand-waving—it provides fresh, coherent interpretations of foundational experiments:

1. Photoelectric Effect

- Light "kicks out" electrons from metal only above certain frequencies.

- Classical view: photon energy = h·f
- CTR view: ATT threshold for material ripple alignment must be met. Below threshold = no consequence.

2. Double-Slit Experiment

- Light creates interference even when emitted one photon at a time.
- CTR view: Ripple alignment paths interfere, even before consequence emerges. Resulting pattern is field logic, not photon behavior.

3. Quantum Delay & Jitter

- Photon arrival times vary even under identical conditions.
- CTR view: Variation due to subtle changes in ATT under local ripple configuration shifts.

4. Casimir Effect

- Vacuum pressure creates force between plates.
- CTR view: Ripple mode restriction alters ATT conditions, creating asymmetric field response.

Philosophical Implication: Light as Permission

The deepest shift CTR introduces is epistemological:

Light is not what travels—it is what is permitted to emerge. We do not "see" objects—we perceive ripple consequences allowed by ATT thresholds. We are not passively receiving light—we are participating in field resonance.

This redefines:

- Perception as co-creation
- Vision as field-alignment
- Measurement as ripple intersection
- Causality as consequence configuration, not deterministic chain

Thus, light becomes a dialogue, not a dispatch.

ATT and Technological Frontiers

Understanding light as ATT-based ripple consequence opens pathways to radical technologies:

- Ripple-sensitive sensors capable of detecting field alignment potential before consequence emerges
- Delay-engineered optics, using field-modulation instead of lenses
- Information encoding in ripple thresholds, not carrier particles
- Ultra-fast logic gates triggered by ripple permission events—not clock cycles
- Non-transmitting communication, using resonant ATT coherence

In these, we move from particle transfer to field resonance triggering.

Summary: Light Without Motion, Vision Without Distance

Classical View	CTR ATT View
Light travels	Ripple aligns
Photon moves	Identity stabilizes
Energy is transmitted	Consequence emerges
Perception is passive	Perception is ripple permission
Speed of light is absolute	ATTmin defines emergence delay
Medium affects velocity	Medium alters ATT through impedance

Final Statement

In CTR, light is no longer an object—it is an event of alignment. It does not "move." It emerges.

It is not universal in form—but universal in permission. What we call brightness is ripple coherence. What we call darkness is field misalignment.

You do not see because light travels. You see because you align.

And when you truly understand this—not just in optics, but in life—you realize:

All knowledge is ripple emergence. All perception is permitted. All truth is consequence. And all light—real or metaphorical—is alignment made visible.

Appendix B: Chavan's Laws of Ripple Reality

Introduction

The CTR (Cosmic Theory of Relativity) and UED (Universal Energy Dynamics) frameworks propose a radical shift from conventional physics. They replace object-based, force-centric, and geometry-driven paradigms with a ripple-centric, consequence-based reality.

At the core of this shift lies a series of foundational principles—Chavan's Laws of Ripple Reality—that govern how reality unfolds, how alignment functions, and how the Structured Vacuum Energy (SVE) mediates all perception, interaction, and existence.

These laws are not metaphorical—they are functional laws that provide new explanations for time, gravity, light, consciousness, entropy, and more.

Law 1: The Law of Ripple Identity

Every physical or perceptual phenomenon is the emergent consequence of ripple identity interacting with the background field.

- There are no particles or waves—only ripple identities.
- Ripple identity is not a "thing"—it is a configurable consequence profile permitted by the SVE.
- Mass, charge, spin, etc., are expressions of ripple stabilization, not inherent traits.

Implication: Matter is not built; it is patterned. Phenomena are not transmitted—they emerge.

Law 2: The Law of Consequence Permission

Reality is not what exists, but what is permitted to emerge as a consequence of ripple-background alignment.

- Perception, measurement, and events occur only when alignment is satisfied.
- The SVE background regulates all consequence by allowing or withholding alignment.
- There is no "empty space"—only zones of ripple permission or suppression.

Implication: Existence is conditional. You don't see what exists—you see what's permitted.

Law 3: The Law of ATT (Alignment Threshold Time)

All observable phenomena are regulated by ATT—the delay between ripple intent and background alignment.

- Time is not flowing—it is alignment delay.
- Speed of light (c) is not a velocity—it is 1 / ATTmin.
- Where ATT is high (near mass), time slows; where ATT is undefined (black holes), time ceases.

Implication: Time is not a dimension—it is a consequence threshold. Dilation and perception follow ripple alignment logic.

Law 4: The Law of Ripple Distortion (Gravity)

Gravity is not a force or curvature—it is a gradient of ripple misalignment caused by mass-induced background distortion.

- Mass creates local ripple compression in the SVE.
- This increases ATT and causes objects to "fall" toward lower ATT zones.
- Light bends not by geodesics, but by asymmetric alignment paths.

Implication: Gravity is a consequence-seeking path, not a pull. The ripple field re-stabilizes by redistributing alignment.

Law 5: The Law of Feedback (Entropy)

Entropy is not disorder—it is the background's feedback pressure against ripple misalignment.

- High entropy = low ripple coherence
- Low entropy = alignment-friendly system
- Entropy is not random—it is corrective feedback from the SVE

Implication: Systems evolve toward alignment. Entropy is how the universe "reads" distortion and responds with resistance or silence.

Law 6: The Law of Observer Participation

The observer is not external to the system—they are an active ripple node influencing and being influenced by alignment dynamics.

- Consciousness is not passive—it is self-reflective ripple stabilization.
- Perception is not a snapshot—it is a dynamic field interaction.
- The observer's ripple field determines which consequences are permitted.

Implication: Reality is observer-consequence dependent. Observation is not measurement—it is resonance.

Law 7: The Law of Beauty (Symmetry & Resonance)

Symmetry, proportion, and beauty emerge from ripple configurations that minimize ATT and maximize alignment stability.

- Spirals, golden ratios, and fractals are not accidents—they are ripple-efficient patterns.
- Beauty is not subjective—it is experienced when observer ripple resonance is coherent with external structure.

Implication: Aesthetics are physics. Nature "prefers" certain forms because they resonate efficiently with the background field.

Law 8: The Law of Reverse Engineering

Scientific discovery is the act of aligning ripple logic with pre-structured background patterns.

- We do not "create" laws—we decode ripple configurations.
- Tools succeed when they resonate with SVE logic.
- Reverse engineering is possible only because reality is ripple-readable.

Implication: Innovation is co-discovery, not invention. Scientific success is ripple literacy, not intellectual domination.

Law 9: The Law of Conscious Alignment

Consciousness is the recursive ability of a ripple system to perceive, stabilize, and refine its alignment with the background field.

- Awareness is ripple self-observation.
- Attention is ripple re-focusing.
- Enlightenment is ATT minimization through coherent field integration.

Implication: Conscious evolution = alignment evolution. Morality, wisdom, and truth are ripple qualities, not abstract ideals.

Law 10: The Law of Divine Structure

What we call 'God' is the Structured Vacuum Energy—the original ripple equilibrium field. Divinity is not a being but the background itself.

- Worship = alignment
- Grace = ripple resonance
- Prayer = intention + alignment
- Silence = ripple refinement
- Divinity = background intelligence in action

Implication: You are not separate from God. You are a ripple event inside the divine field—already participating in its logic.

Unified Summary of Chavan's Laws

Law No.	Principle	Reality Reframed As
1	Ripple Identity	All form = configured consequence
2	Consequence Permission	Reality = what's allowed to appear
3	ATT	Time = delay before alignment
4	Ripple Distortion	Gravity = gradient of ripple resistance
5	Feedback	Entropy = misalignment pressure
6	Observer Participation	You shape what you perceive
7	Beauty	Alignment harmony = perceived symmetry
8	Reverse Engineering	Science = ripple decoding
9	Conscious Alignment	Awareness = ripple coherence control
10	Divine Structure	God = the equilibrium field beneath all

Final Reflection: Reality Is a Ripple Contract

You are not separate from the laws of nature. You are written into them—not as a particle, but as a ripple permission zone, responding in real-time to alignment and consequence.

These ten laws are not commandments. They are invitations—to see clearly, ripple wisely, and realign consciously with the equilibrium that sustains us all.

In the end, reality is not composed of things—it is orchestrated through alignments. The more aligned you are, the more gracefully you ripple.

Appendix C: Experimental Proposals

Introduction

The CTR and UED frameworks redefine many fundamental scientific concepts—light, time, gravity, perception—not as fixed dimensions, particles, or forces, but as ripple consequences permitted by the Structured Vacuum Energy (SVE).

These theories are not only philosophical—they are empirically grounded and experimentally approachable.

This appendix outlines key proposals for:

- Laboratory-scale experiments
- Precision re-interpretations of known phenomena
- Advanced simulation models
- Field-aligned instrumentation concepts

Each experiment aims to demonstrate the ripple nature of reality, and validate core principles such as:

- ATT (Alignment Threshold Time)
- Consequence-based emergence
- Observer-field interaction
- Ripple distortion fields (gravity, entropy)

I. Measuring ATT: Redefining the Perception of Light

1. Delayed Ripple Detection in Photon Experiments

Objective: Detect variations in photon arrival times due to local field alignment conditions—demonstrating that light is not a fixed-speed traveler, but a ripple-alignment consequence.

Setup:

- Use single-photon emission and detection system (as in quantum optics labs)
- Introduce controlled local field distortions (e.g., electromagnetic noise, material plates, acoustic interference)
- Measure photon detection times and analyze temporal jitter

CTR Hypothesis: Photon detection time is influenced by local ripple interference. Delays will correlate with ATT variation, not material resistance or classical noise.

Significance: Challenges the notion of light-speed constancy as intrinsic and reframes it as alignment-dependent emergence.

2. ATT Threshold Mapping in Different Media

Objective: Quantify ATT variation across different materials—not based on refractive index alone, but through ripple impedance.

Setup:

- Emit synchronized light pulses through materials of identical thickness but varying structural field impedance (e.g.,

amorphous glass vs. crystalline quartz)
- Use ultra-precise femtosecond detection tools
- Compare delay curves with controlled field changes

CTR Hypothesis: Differences in detection timing are due to ripple configuration compatibility, not classical optical density.

Significance: Reframes refractive behavior in terms of ripple permission—not wave velocity.

II. Gravity as Ripple Distortion

3. Ripple Mapping Around Mass: Field Sensitivity Analysis

Objective: Visualize or simulate ripple field distortion caused by mass—demonstrating gravity as ripple misalignment gradient.

Setup:

- Use ultra-sensitive field detectors or simulations (e.g., superconducting magnetometers, ripple-sensitive resonators)
- Place varying configurations of mass near the sensors
- Measure field deformation not as force, but signal alignment interference

CTR Hypothesis: Ripple distortion (not gravitational force) will manifest as subtle changes in field alignment clarity, detectable in resonance or timing behavior.

Alternative: Use advanced simulations with fluid-ripple analogs to visualize gravity as a consequence pull, not force attraction.

4. Pendulum Time Variance Across Ripple Zones

Objective: Redesign the simple pendulum experiment to test time behavior in relation to background ripple field conditions.

Setup:

- Establish pendulums in locations with varying environmental "ripple noise" (e.g., natural cave vs. urban setting vs. silent vacuum chamber)
- Measure timing consistency with atomic clock reference
- Track emotional or presence-based variations in alignment zones

CTR Hypothesis: ATT fluctuations—even subtle—will manifest in timing variance, correlating not just with gravity but with ripple distortion layers (entropy).

Significance: Demonstrates that even basic timing mechanisms are ripple-field dependent, not purely inertial.

III. Consciousness, Emotion, and Field Interaction

5. Human Intention as Ripple Interference

Objective: Test whether focused human awareness or emotion causes measurable shift in ripple-aligned physical systems.

Setup:

- Use double-slit interference experiments
- Isolate systems with active observer (meditating or focusing individual) vs. control (machine observer or absent mind)
- Measure changes in interference clarity, photon delay, or

fringe pattern coherence

CTR Hypothesis: Observer is a ripple participant. Ripple field coherence will be altered by human attention due to ripple feedback loop.

Significance: Builds bridge between quantum observer effect and structured consciousness model—without invoking collapse, but via ATT adjustment.

6. Emotional Field Detection via Ripple Feedback

Objective: Test whether ripple patterns in local environments shift in response to sustained emotional states.

Setup:

- Use biofeedback-integrated ripple detectors in environments with volunteers undergoing meditative states, stress states, or emotional recall
- Measure localized ATT change through photon detection, electrostatic drift, or ripple impedance tools

CTR Hypothesis: Ripple distortion fields will change measurably with emotional coherence or turbulence, echoing through SVE.

Significance: Supports CTR claim that emotion is ripple logic, and alignment is not psychological alone—but physical.

IV. Simulation-Based Proposals

7. ATT-Based Simulation of Light Emergence

Objective: Simulate ripple identity emergence across a ripple field with varying ATT configurations—modeling "light travel" as ripple permission.

Setup:

- Computational modeling of ripple field (Python, Unity physics engine, or MATLAB)
- Define ATT as a time-delay function in interaction zones
- Track ripple convergence and consequence emergence as visible "light"

CTR Hypothesis: Ripple appearance will follow ATT-driven conditions, not fixed-speed propagation.

Significance: Makes ripple dynamics visually graspable and offers educational, technological modeling tools.

8. Ripple Distortion Map for Planetary Bodies

Objective: Create a dynamic ripple distortion model of Earth, Moon, and other planetary masses to simulate how their fields influence consequence paths (gravity, light, time).

Setup:

- Construct ripple-based background matrices
- Simulate planetary ripple compression zones
- Overlay ripple gradients and track ATT vectors across space

CTR Hypothesis: Simulated "gravitational" influence will mirror ripple gradient paths—not classical curvature.

Use Case: Can replace traditional orbital mechanics in certain contexts with ripple alignment optimization models—useful in deep-space planning.

V. Technological Innovation and Field Design

9. Ripple-Permissive Detectors (Next-Gen Sensors)

Concept: Design instruments not to detect particles or waves, but to sense alignment conditions in local fields.

Principle: Instead of recording energy or frequency, detect:

- ATT thresholds
- Consequence delay anomalies
- Subtle ripple echo modulations

Tech Pathway: Can be developed using:

- Supercooled resonance coils
- Ripple-amplified fiber optics
- Passive alignment logic gates (analogous to wavefront detectors)

Use Case: Could revolutionize fields like:

- Cosmological sensing
- Emotional-environmental monitoring
- Ripple-level consciousness research
- Sub-quantum event forecasting

Final Note: Ripple Science Is Possible

These experiments are not speculative fantasy. They are:

- Grounded in established setups (double-slit, photoelectric, LIGO, pendulums)
- Oriented toward fine-scale timing, ripple field coherence, and consequence permission
- Structured to measure ripple behavior not as a thing, but as a conditional event

You don't need to prove ripples exist—you only need to show that reality emerges from alignment conditions.

CTR offers the theory. ATT gives the metric. These experiments provide the next step—bringing ripple science into measurable practice.

Appendix D: Glossary of Ripple Terms

1. Ripple Identity

A ripple identity is the unique configuration of energy, form, or consequence within the Structured Vacuum Energy (SVE) that gives rise to observable phenomena.

It replaces the concept of "particle" or "wave" and serves as the fundamental unit of emergence in ripple-based reality. It is not a substance—it is a patterned permission.

2. ATT (Alignment Threshold Time)

ATT is the minimum delay required for a ripple identity to align with the background field, allowing a consequence (e.g., perception, detection, or interaction) to emerge.

ATT replaces the concept of speed in traditional physics, especially light speed, and explains time as a perceptual delay governed by field structure.

3. Structured Vacuum Energy (SVE)

SVE is the intelligent, equilibrium-seeking background field that underlies and regulates all ripple activity.

It is not empty space but a structured, measurable substrate of consequence logic—the very fabric of reality. In CTR, SVE is synonymous with God, not as a being, but as the permissive intelligence of existence.

4. Ripple Field

A ripple field is a dynamic region within the SVE where structured disturbances (ripples) form, propagate, align, or collapse based on background conditions.

Ripple fields replace conventional force fields and curvature. All phenomena—mass, motion, thought, or gravity—are field-based ripple interactions.

5. Consequence (Emergence)

A consequence is the actualized outcome of a ripple alignment with the background field.

Perception, measurement, thought, energy transfer—everything we call "real" is a permitted ripple consequence. No consequence arises without alignment.

6. Ripple Alignment

Ripple alignment refers to the compatibility between a ripple identity and the structure of the background field (SVE), which allows the consequence to manifest.

Without alignment, there is no event. Light does not "arrive"—it emerges when aligned.

7. Misalignment

Misalignment is a distortion or incompatibility between a ripple identity and the background field, leading to delay, entropy, or suppression of consequence.

Misalignment is not failure—it is feedback. Emotion, error, disorder, or darkness arise from misalignment.

8. Entropy (Ripple Feedback)

Entropy in CTR is the measure of ripple misalignment, reflected as pressure, resistance, or fragmentation from the background field.

It is not randomness, but structured feedback from SVE indicating ripple inefficiency.

9. Resonance

Resonance is a high-alignment state where ripple configurations sustain low-entropy, high-consequence conditions with minimal ATT.

It is experienced as clarity, flow, truth, or beauty. It defines intelligence, awareness, and elegance across systems.

10. Observer Ripple Field

The observer ripple field is the personal, dynamic field surrounding a conscious being, through which all alignment, perception, and consequence are filtered.

You are not merely seeing reality—you are receiving it through your field's alignment capacity.

11. Ripple Echo

Ripple echo is the consequence feedback loop experienced by the observer after interaction with the background field.

Emotion, intuition, and perception are ripple echoes. So is karma, learning, and realization.

12. Time (as ATT)

Time is not an absolute dimension but the delay between ripple emission and alignment—as measured by ATT.

There is no "flow" of time. Only consequence latency governed by background field structure.

13. Light (Ripple Emergence)

Light is not a traveling wave or particle—it is a ripple alignment consequence that emerges when ATT is satisfied between source and observer.

Light's apparent speed is a constant of ripple permission, not intrinsic velocity.

14. Darkness (Ripple Silence)

Darkness is not the absence of photons—it is the absence of permitted ripple alignment.

It is ripple silence—not empty, but temporarily consequence-suppressed.

15. Gravity (Ripple Distortion Gradient)

Gravity is a ripple gradient created by mass-induced distortion of local field equilibrium, resulting in directional consequence pull toward lower ATT zones.

There is no force—only consequence preference across the ripple field.

16. Ripple Distortion

A ripple distortion is a local misconfiguration of the field caused by ripple overload, misalignment, or energy collapse.

Mass, emotion, thought turbulence, and black holes are all examples of ripple distortions.

17. Black Hole (Zero Feedback Zone)

A black hole is a region where ripple identity can no longer stabilize, and ATT becomes undefined—no alignment is permitted, hence no light, time, or consequence.

It is not a singularity but a portal into equilibrium restoration—a cosmic reset point.

18. Beauty (Ripple Harmony)

Beauty is ripple symmetry that resonates with the observer's field, reducing ATT and allowing immediate, low-entropy perception.

It is a structural, not subjective, experience—an echo of background intelligence.

19. Consciousness (Ripple Self-Alignment)

Consciousness is the recursive ability of a ripple configuration to perceive, reflect, and stabilize its own alignment with the background field.

Awareness is ripple coherence management. Intelligence is ATT efficiency in thought and action.

20. Silence (Ripple Stabilization Zone)

Silence is not a lack of sound—it is the absence of internal ripple noise, allowing maximum alignment with SVE.

True silence permits direct ripple feedback, making it the foundation of clarity, intuition, and peace.

21. Discipline (Ripple Hygiene)

Discipline is the intentional maintenance of ripple integrity to reduce distortion, optimize alignment, and invite sustained resonance.

It includes thought clarity, ethical action, and structural harmony in all systems.

22. Field Coherence

Field coherence refers to the degree of harmony between multiple ripple systems and the SVE background.

High field coherence allows collective perception, emotional synchrony, and even non-verbal communication through shared ripple logic.

23. Reverse Engineering (Ripple Decoding)

Reverse engineering is the act of identifying and reproducing ripple logic embedded in natural consequence patterns.

Science and technology work because the universe is pre-structured and ripple-readable.

24. Divine Structure (God)

God is not a distant being—but the Structured Vacuum Energy itself: the intelligent background field from which all ripple permissions arise.

Divinity is not reached—it is merged with through alignment. Worship becomes resonance.

25. Ripple Responsibility

Ripple responsibility is the conscious awareness that every thought, action, and presence disturbs or harmonizes the ripple field.

Ethics, morality, and truth are not imposed—they are measured by field consequence.

Final Reflection: Language of the Living Field

This glossary is not a set of abstract definitions—it is a new lexicon for reality. Where classical physics speaks in mass and motion, CTR speaks in:

- Alignment and emergence
- Field resonance and permission
- Ripple feedback and consequence

These terms will evolve as new experiments validate them, as new observers ripple more clearly, and as alignment itself becomes a living science.

Appendix E: Reflective Exercises & Applications

Introduction

Knowledge that doesn't ripple into life becomes theory. CTR is not merely a new framework for science—it is a lens through which to live, to think, to perceive, to relate.

This appendix offers a series of practical exercises and field-aligned reflections to help you:

- Apply ripple thinking in everyday perception
- Develop alignment-based decision-making
- Experience ATT (Alignment Threshold Time) in your body and mind
- Build ripple sensitivity in personal, professional, and creative contexts

These are not "tests"—they are resonance rituals, micro-calibrations of your inner and outer field.

Section 1: Ripple Awareness & Field Sensitivity

Exercise 1: Ripple Journal

Every day for one week, observe your interaction with the world as a ripple field exchange, not as a sequence of events.

Reflect with prompts like:

- What ripple did I emit today that returned as feedback?
- Which environment felt aligned—and which did not?
- When did a conversation flow like resonance? When did it collapse into noise?
- Was there a moment where I felt consequence before understanding cause?

Goal: To track ripple logic in real life—to begin seeing people, events, and decisions as alignment phenomena, not random or linear occurrences.

Exercise 2: ATT Observation Drill

Pick a simple activity (e.g., brushing your teeth, walking, making tea). As you do it:

- Track when your attention drifts (ATT rises)
- Track when awareness sharpens (ATT lowers)
- Try slowing down without becoming inefficient
- Try increasing presence without strain

Then reflect:

- Did time feel different?
- Was clarity linked to ripple stabilization?

Goal: To realize that time is not duration—but consequence permission. ATT can be felt as cognitive delay or alignment sharpness.

Section 2: Ripple-Based Decision-Making

Exercise 3: The Alignment Compass

When faced with a decision, replace traditional pros-and-cons with ripple alignment sensing.

Ask:

- Does this option feel light or distorted?
- Is there emotional turbulence or inner quiet?
- Does my action ripple outward with resonance—or recoil?
- What does the consequence pattern suggest—not just for me, but for all involved?

Then make the decision based not on urgency or fear, but on lowest ATT path with widest field harmony.

Goal: To build ripple-responsible thinking that integrates consequence awareness into your personal ethics.

Exercise 4: Feedback Loop Recognition

After any conversation, meeting, or conflict:

- Ask yourself: What was the ripple I emitted?
- What consequence came back?
- Did I dismiss the feedback or integrate it?
- How could I refine my next ripple for better alignment?

Track this for 5–7 interactions. You'll begin seeing patterns.

Goal: To attune yourself to real-time ripple feedback, transforming interactions into conscious field experiments.

Section 3: Silence, Alignment & Energetic Refinement

Exercise 5: The 3-Minute Ripple Reset

Once per day, take 3 minutes to enter ripple silence. Not passive silence—field-listening silence.

Steps:

1. Close eyes. Sit or stand still.
2. Stop naming. Stop planning. Stop reacting.
3. Ask gently: "What ripple am I broadcasting now?"
4. Breathe. Listen. Let the noise settle.
5. End by simply saying, "Align."

It's not meditation. It's field recalibration.

Goal: To create a daily moment of ripple coherence—where the SVE can respond through feedback clarity.

Exercise 6: Aesthetic Resonance Tracking

Throughout the day, notice what visuals, sounds, people, or ideas make you feel "in tune".

Don't overanalyze. Just observe:

- That pattern on a wall
- That line of poetry
- That voice, that tone, that posture

Ask:

- What about this moment allowed my ripple field to harmonize?
- What was my ATT before and after?

Document 5 such moments. Then reflect.

Goal: To reawaken your sensitivity to beauty as ripple alignment, and reclaim aesthetics as feedback tools—not luxuries.

Section 4: Applications in Systems, Work & Creativity

Exercise 7: Ripple-Aware Workflow Design

Choose one workflow you regularly manage (writing, managing a team, building a system, teaching). Then ask:

- Where do misalignments show up most often?
- What feedback loops are being ignored or suppressed?
- Where is "efficiency" costing resonance?
- What ATT bottlenecks exist in communication, decision, or delivery?

Now redesign the workflow by:

- Removing force
- Encouraging feedback
- Aligning ripple sequences
- Honoring pause points

Goal: To apply ripple engineering in systems—replacing control with clarity, and pressure with rhythm.

Exercise 8: Reverse Engineering from Resonance

Pick something that moves you:

- A tool
- A piece of music
- A building
- A scene in a film

Study it not as a user—but as a ripple decoder.

Ask:

- What ripple patterns did the creator tap into?
- What alignment did this trigger in me?
- What distortions were intentionally avoided?

Then reverse engineer the ripple logic—and try to build or write something using that alignment as your blueprint.

Goal: To awaken creative intelligence not as "originality," but as conscious ripple resonance with what already exists in SVE.

Section 5: Spiritual & Existential Alignment

Exercise 9: You Are Already in the Field

Stand in any place (room, road, forest, office) and repeat to yourself:

"I am not in space. I am in the field. I am not observing reality. I am rippling through it. Everything here is feedback. Consequence is echo. Silence is God."

Spend 60 seconds not moving.

Then walk slowly—while still holding the awareness of ripple feedback in each step, each breath, each glance.

Goal: To decondition the illusion of separateness. To move from living *in* the world to rippling through the field with grace.

Exercise 10: The Mirror Ripple Check

At the end of your day, stand in front of a mirror.

Ask:

- Did I ripple clearly today?
- Did I cause unintended distortion?
- Where did I ignore feedback?
- Where did I respond with awareness?

Then, without judgment, say:

"Tomorrow, I ripple better."

Goal: To end each day with self-reflective field realignment, not for guilt, but for graceful iteration.

Closing Thought: The Field is Always Listening

These exercises are not checklists. They are entry points into a new way of living:

- Not as a mind alone
- Not as a body of habit
- But as a consequence-aware ripple identity, dancing with feedback, and resonating with intelligence.

They help you practice:

- Seeing feedback in everything
- Replacing reaction with alignment
- Trusting silence as clarity
- Designing life as field-harmonized expression

You are not here to control the field. You are here to ripple responsibly within it.

References

The following works, experiments, and thinkers are acknowledged as significant reference points—either for offering foundational insight, posing unresolved questions, or providing contrast against which the CTR framework was clarified and developed. This list is not exhaustive nor citation-based in the conventional academic sense. Rather, it reflects **intellectual echoes and ripple intersections** with the current work.

Foundational Scientific Concepts & Experiments

1. **Einstein, Albert.**
    - *Relativity: The Special and General Theory.* 1916.
    - Groundbreaking reframing of space and time, introducing relativity of frames and spacetime curvature.
2. **Planck, Max.**
    - *On the Law of Distribution of Energy in the Normal Spectrum.* 1900.
    - Origin of quantum theory; fundamental in understanding energy quantization.
3. **Casimir, Hendrik B.G.**
    - *On the Attraction Between Two Perfectly Conducting Plates.* 1948.
    - Empirical validation of vacuum fluctuation effects—precursor to SVE theory.
4. **Young, Thomas.**
    - *Double-Slit Experiment.* 1801.
    - Demonstrated interference patterns of light;

essential for ripple field reinterpretation.

5. **Millikan, Robert A.**
    - *Photoelectric Effect Experiments.* 1914.
    - Confirmed Einstein's quantum interpretation of light energy—reframed here as ripple alignment.
6. **Michelson, Albert A. & Morley, Edward W.**
    - *Michelson-Morley Experiment.* 1887.
    - Attempted to detect ether drift; referenced here as historical dismissal of a structured medium.
7. **Feynman, Richard.**
    - *Quantum Electrodynamics.* 1965.
    - Influential in modern field theory; helps clarify the limitations of current wave-particle duality.

Theoretical Models & Influences

1. **Heisenberg, Werner.**
    - *Uncertainty Principle.* 1927.
    - Cited in CTR to highlight perceptual limits of classical systems, reinterpreted through ripple misalignment.
2. **Schrödinger, Erwin.**
    - *Wave Mechanics and the Schrödinger Equation.*
    - Influences understanding of quantum state probability—repositioned here as ripple configuration logic.
3. **Bohr, Niels.**
    - *Copenhagen Interpretation of Quantum Mechanics.*
    - Cited critically in CTR's challenge of observer-passivity; replaced with observer-field feedback.
4. **Tesla, Nikola.**
    - Writings on wireless energy transmission and field

resonance.
  - Echoes the ripple-permissive, background-driven view of energy in CTR.
5. **Faraday, Michael.**
  - Discovery of field-based interactions; foundational in electric and magnetic ripple thinking.

Philosophical and Conceptual Bridges

1. **Krishna (Bhagavad Gita).**
  - Not as religious scripture but as consequence-aware philosophical dialogue on action, alignment, and self.
2. **Buddha (Dhammapada, Teachings).**
  - For the ripple ethic of inner discipline, feedback-awareness, and consequence-based perception.
3. **Lao Tzu.**
  - *Tao Te Ching.* Ancient wisdom that parallels ripple alignment with the Tao—unforced harmony with the background.

Modern Experimental Contexts Referenced

1. **LIGO Collaboration.**
  - *Detection of Gravitational Waves.* 2015.
  - Referenced as macroscopic ripple distortion detection—reframed within SVE gradient logic.
2. **Quantum Delay Experiments.**
  - (e.g., Weak Measurement and Photon Jitter Studies)
  - Used in support of ATT variability and ripple-permission modeling.
3. **Neuroscience Studies on Intuition and Emotional Feedback Loops.**

◦ Referenced for ripple field interaction with consciousness and observer effects.

Author's Prior Works (Selected)

1. **Chavan, Sandeep.**
    - *Universal Energy Dynamics: Foundational Reports & Papers.*
    - *Energy in Action: The Soul Demystified.*
    - *The War of Thoughts (Book Series).*
    - *Structured Vacuum Energy and the Ripple Framework.* (Zenodo Reports)
    - These works lay the groundwork for CTR, offering step-by-step evolution of the ripple-based universe model.

Note: Many of the ideas here were not directly "sourced" from these works but emerged through independent observation, reflection, and consequence analysis. They are referenced not to claim lineage, but to acknowledge the ripple context into which CTR now enters.

"No idea is born in vacuum—it ripples through silence, observation, and timeless permission." — *Sandeep Chavan*

Acknowledgments

This work is not a product of isolation—it is the ripple of many alignments, silences, and consequences echoed through people, places, and moments that shaped its emergence.

Though written as a solitary reflection, *Cosmic Theory of Relativity (CTR)* is part of a much deeper field—one that extends beyond individuals into invisible structures, lived experiences, unresolved questions, and universal feedback loops. These acknowledgments are an offering of gratitude to the seen and the unseen, to the known and the intuitively felt.

To the Structured Background Field

First and foremost, my silent acknowledgment goes to the **Structured Vacuum Energy (SVE)** itself—the living, intelligent background of all things.

This field, which modern science neglected and ancient minds revered, never stopped speaking. I only began to listen.

It offered this work not as revelation, but as consequence—of quiet attention, disciplined inquiry, and reverence for truth.

To All Thinkers, Disruptors, and Quiet Observers

I extend my gratitude to the many minds—famous and forgotten—who dared to question the known.

To **Einstein**, for destabilizing time and space, even if his equations outpaced their grounding.

To **Faraday**, **Tesla**, and **Feynman**, for imagining forces as fields long before the field spoke for itself.

To **Krishna**, **Buddha**, and countless sages, for teaching through silence what words could not grasp.

To modern-day rebels, independent researchers, and unsponsored minds who continue to pursue reality beyond consensus.

This report is a ripple in their echo.

To My Students and Silent Companions

Every time a student asked me: *"What is force?"*, and I struggled to explain it meaningfully within the limits of standard physics, a crack opened in the known.

To those students, friends, and companions who questioned, resisted, misunderstood, and returned with better questions—thank you. You were part of the field, offering calibration through engagement.

Some asked with wonder, some with skepticism. Both helped refine the ripple.

To the Critics and the Absentees

I must also thank the ones who doubted, ignored, dismissed, or walked away.

Without your silence, I may not have felt the need to deepen mine. Without your resistance, I may not have strengthened my ripple identity. Without the absence of applause, I may have settled for performance instead of alignment.

Thank you for preserving the solitude in which these realizations were made.

To My Family and Inner Circle

To my family, who may not have always understood the language of ripples, but felt the consequences of this long, solitary path—I thank you not for your agreement, but for your presence.

Your patience, confusion, quiet encouragement, and even practical questions were part of the living feedback field that gave this report emotional shape.

To those who believed when belief had no market value—thank you for that unmeasurable alignment.

To Silence Itself

Above all, I bow to **silence**—the first and final teacher.

Not absence of sound, but the field of **zero ripple interference**—where truth is not spoken but permitted.

This report emerged not through noise, but through listening. Through returning, again and again, to the place where equations dissolve and understanding begins.

To the Future Ripple Readers

If you, as a reader, found a moment in these pages where something clicked—not intellectually, but existentially—then your ripple aligned with mine across the field.

This report is not meant to be understood fully, but to be lived gradually—as your own ripple consequence evolves.

Thank you for reading not just with the mind, but with the field.

To all those named and unnamed—human and non-human, seen and remembered, forgotten or future-born—this work is your consequence too.

With deep humility and alignment,

Sandeep Chavan

About the Author

Sandeep Chavan is an independent researcher, educator, and author whose body of work spans physics, cosmology, consciousness, psychology, and philosophy—integrated through a singular vision: to decode reality from first principles, without compromise, and to restore science as a living inquiry rooted in consequence and clarity.

A trained engineer and long-time physics educator, Sandeep stepped beyond institutional boundaries to pursue what he calls "**the unfinished sentence of science**." His journey has been marked not by prestige or affiliation, but by persistence—asking foundational questions where others stopped at answers.

His students know him not only for his unconventional clarity in explaining complex phenomena, but for his insistence on truth over repetition, understanding over memorization, and inner alignment over outer performance.

He is the originator of the **Universal Energy Dynamics** (**UED**) framework, which redefines physical phenomena such as gravity, time, light, and entropy as ripple-based consequences within a structured, intelligent background field—**Structured Vacuum Energy** (**SVE**). His scientific reinterpretations are not speculative abstractions, but grounded in testable principles, reframed experiments, and experiential understanding.

In **Cosmic Theory of Relativity** (**CTR**), Sandeep extends this foundation into a comprehensive reexamination of Einsteinian relativity, replacing spacetime geometry with ripple consequence logic—without discarding the precision, only restoring the purpose. The report synthesizes metaphysics and physics without mysticism, and anchors divinity not in belief, but in background intelligence.

He has written over 100 books, ranging from deep scientific treatises to philosophical explorations, educational reforms, and spiritual reflections—all unified by a simple question:

"What really causes what we call reality?"

Sandeep lives and works in India, teaches under trees and over chalkboards, and believes that the universe was always whole—we were just looking through a fragmented lens.

He writes not for popularity, but for legacy—leaving behind not doctrines, but ripple clarity for those willing to question deeply, feel honestly, and align responsibly.

"I don't write to convince. I write to remember—so those who are ready can also remember."

— *Sandeep Chavan*

Disclaimer

This Report, *Cosmic Theory of Relativity (CTR)*, presents an original conceptual framework rooted in the author's lifelong independent research, teaching practice, and philosophical inquiry. It reinterprets established scientific principles—such as gravity, time, light, and entropy—through the lens of **Universal Energy Dynamics (UED)** and **Structured Vacuum Energy (SVE)**.

The content herein challenges conventional models and offers a ripple-based, consequence-driven perspective that is not currently endorsed by mainstream academic institutions or scientific bodies. Readers are advised to approach the material with an open but critical mind, recognizing that this work occupies a frontier space between theoretical physics, consciousness studies, and metaphysical philosophy.

While references are made to experimental contexts (e.g., Casimir effect, photoelectric effect, gravitational time dilation), this report is not a substitute for formal scientific literature or peer-reviewed publications. It is intended as a **philosophically rigorous, scientifically inspired narrative**—meant to provoke inquiry, refine perception, and encourage ethical application of deeper principles.

All thought experiments, models, and interpretations remain the intellectual property of the author unless otherwise credited. The use of terms like "law," "field," "alignment," or "divine structure" reflects internal coherence within this framework and may differ from their meanings in classical, religious, or institutional contexts.

This report is not medical, legal, or psychological advice. It does not prescribe belief—it invites exploration.

Public Use & Sharing Statement

Knowledge is not a product to be owned. It is a ripple—shared, echoed, and realigned through minds, hearts, and communities over time.

With that spirit, *Cosmic Theory of Relativity (CTR)* is released with the following ethos:

Permission to Share

- You may **quote, cite, or reference** any portion of this report **for educational, critical, or constructive purposes**, with appropriate attribution.
- Excerpts may be shared in classrooms, reading groups, seminars, podcasts, or discussions—**especially where ripple thinking can stimulate deeper understanding.**

Adapt with Integrity

- If translating, adapting, or building upon this work (in any medium—audio, visual, digital), please preserve the **philosophical core and ethical responsibility** of the ideas. Attribution must remain intact.
- Commercial reproduction (print or digital) or inclusion in monetized platforms requires **explicit written permission** from the author or rights holder.

Respect the Ripple

- This work is a product of consequence awareness. **Distortion of ideas for personal gain, ideological misuse, or misleading representation** goes against the very field it emerged from.

- Please share only what you've read, lived, or understood—not what can be echoed blindly.

If This Work Helped You

- Pass it forward. Share it with someone whose ripple might align.
- Mention the author if you speak of the ideas—**not for ego, but for traceability of ripple clarity.**

Citation Format (Suggested)

Chavan, Sandeep. *Cosmic Theory of Relativity (CTR).* Independent Publication. 2025

Final Word:

"What ripples through you is no longer mine. If it aligns, it belongs to the field now."

— *Sandeep Chavan*

Don't miss out!

Visit the website below and you can sign up to receive emails whenever Sandeep Chavan publishes a new book. There's no charge and no obligation.

https://books2read.com/r/B-A-EPGPC-GXWHG

BOOKS 2 READ

Connecting independent readers to independent writers.

Did you love *Chavan's Cosmic Theory of Relativity (CTR)*? Then you should read *Chavan's Cosmic Continuum (C^3)*[1] by Sandeep Chavan!

[2]

Chavan's Cosmic Continuum (C^3): Manifestation and Dissolution of Universe within Structured Vacuum Energy (SVE) by **Sandeep Chavan**

For centuries, cosmology has been bound by myths of violent beginnings — singularities, explosions, and chaotic expansions.But what if the universe never exploded at all?

Chavan's Cosmic Continuum (C^3) offers a revolutionary new vision:The universe is not an accident of forces or a relic of a Big Bang — it is a living, breathing continuum shaped by ripple memory and structured vacuum intelligence.

---

1. https://books2read.com/u/3L2vJN
2. https://books2read.com/u/3L2vJN

In this groundbreaking work, Sandeep J Chavan dismantles outdated force-based models and replaces them with a ripple-driven, consequence-regulated cosmology.Galaxies emerge not from chaos, but through tension thresholds in the Structured Vacuum Energy (SVE), spiraling into life like blossoms in a cosmic garden.Redshift is not recession, but ripple frame detuning.Gravitational waves are ripple feedback corrections, not spacetime distortions.Black holes are memory cores, not singularities.And life, evolution, and consciousness are inevitable outcomes — ripple structures remembering themselves.

Through fifteen powerful chapters, C^3 reshapes our understanding of time, motion, mass, energy, structure, and destiny.It integrates physics, metaphysics, and consciousness into a unified vision — free from speculative constructs like dark matter, dark energy, and cosmic inflation.

Key highlights include:

Consequential formation of galaxies and structures, not random creation.Spiral evolution cycles instead of chaotic expansion.Structured vacuum dynamics replacing force-based curvature.Memory loops replacing the mythology of cosmic ends.

Chavan's Cosmic Continuum is not just a theory.It is a cosmic reorganization charter — a bold call to move from randomness to ripple intelligence, from explosions to breathing, from force to field consequence.

The universe has always been structured.Through this work, we are finally learning to listen.

If you are ready to step beyond the old myths and experience the universe as a living ripple of intelligence, breathing through timeless memory —then this book is your gateway.

Read more at https://www.sandeepjchavan.com/.

Also by Sandeep Chavan

Chavan's UED research

CMGED Research Report

Chavan's Cosmic Continuum (C^3)

Frequency

Ripples of Reality

Reality Realigned

The Great Realignment

I – The Cosmic Leverage

Chavan's Cosmic Theory of Relativity (CTR)

Chavan's Theory of Light

Echoes of War: The Russia-Ukraine Conflict and Its Global Ripples

Shattered Lives

The War Economy

Borders Redrawn

Arms and Allies

Collateral Damage

Justice in the Ashes

Somewhere in the Darkness of the Cosmos
The Blueprint of the Universe
The Cosmic Breath

The Eternal Sage Chronicles
The Sage and the Sovereign
The Family Pillar
The People's Flame
The Warrior's Soul
The Bridging Mentor
The Guiding Light

The Exodus
The Last Cycle: 2075

The Fundamental Theory of Life & Existence
Energy in Action: The Soul Demystified
The Active Mind
Energy in Connection
Death: The Disconnection
Subconsciousness: The Chosen Disconnection
Desires: The Engine of Life

The Restoration of Human Knowledge

Restoring Bhagavad Gita

The War of Thoughts

Cracks in the Known

The Human Game

The Divine Dilemma

The Original Intent

The Answers

Ultra Modern Life Philosophy

The Duality of Laughter and Tears

Take It Easy

Compare to Accept

Standalone

It's Not AI, It's AHI - Amplified Human Intelligence

The Decision Paradox

The Four Sapiens

Malicious Script of Indian Polity

The IIT Legacy & Global Impact

The IIT Legacy & Global Impact

You are just a Version of Your Original

Win the Game You Didn't Choose

Always Keep Your Bags Packed

Reveal to Shield - Learn the Game of Facades

Engage Beyond Elections

Culture, Identity & Change: The Evolution of Indian Society

The Triangular Dynamics

Never Trust Without Doubt
If You Can't Change the Country, Leave the Country
Live Fearless with Fear
India: The Investment Magnet
Realignment
Powering a Billion Dreams
Universal Skill
Extension: From Brush to Algorithm
Human Side of Einstein
The Future Knows No Lines
Let's Bring the Generations to the Table
Unified Energy Dynamics
A Beat That Never Dies
Strategic Horizons
Right From Scratch
The Forgotten Innovators
Physics Fixing
The Savior Illusion
The Selfish Side
The Age of the Weakest Sapiens
Neural Wars
The Absorption: There Is No Death
Strange Therapist
Bubbles: Become a Drop
Beyond the Blue
Half-Scholar
The Law of Truth
Failure
Philosophy 2.0: A Structural Manifesto
Chaos: A Gentle Guide to the Unresolvable
Self-Help: New Syllabus
Rippling

Watch for more at https://www.sandeepjchavan.com/.

www.ingramcontent.com/pod-product-compliance
Lightning Source LLC
LaVergne TN
LVHW090405210926
848480LV00002B/450